THE COMING STORM

To my parents, William and Catherine Fox

LIAM FOX

THE COMING STORM

WHY WATER WILL WRITE THE 21ST CENTURY

\B^b\
Biteback Publishing

First published in Great Britain in 2024 by
Biteback Publishing Ltd, London
Copyright © Liam Fox 2024

Liam Fox has asserted his right under the Copyright, Designs and Patents Act 1988 to be
identified as the author of this work.

ISBN 978-1-78590-859-0

10 9 8 7 6 5 4 3 2 1

A CIP catalogue record for this book is available from the British Library.

Set in Minion Pro and Trajan Pro

Printed and bound in Great Britain by
CPI Group (UK) Ltd, Croydon CR0 4YY

FSC
www.fsc.org
MIX
Paper | Supporting
responsible forestry
FSC® C171272

CONTENTS

LIST OF ILLUSTRATIONS AND PLATES

ILLUSTRATIONS

1. Salt composition of the sea. https://www.grida.no/resources/5606 © Philippe Rekacewicz, February 2006.
2. The Ogallala Aquifer. Via Wikimedia Commons, CC BY-SA 3.0.
3. Guarani Aquifer. © Earthbeat a project of *National Catholic Reporter*.
4. The Nubian Sandstone Aquifer. *Sci Rep* 11, 78 (2021). https://doi.org/10.1038/s41598-020-80160-0.
5. The Kingdom of Saudi Arabia. © PSIPW, 2011.
6. The shrinking sea. © W. H. Freeman and Company.
7. Map of Tigris–Euphrates basin. Created by Karl Musser, based on USGS data, Wikipedia, CC BY-SA 2.5.
8. Jordan River basin. © United Nations Environment Programme (UNEP).
9. Major rivers sourced in Tibet. www.meltdownintibet.com © Michael Buckley.
10. Map of Mediterranean Sea and Lake Nasser. By Shannon1, Wikimedia Commons, CC BY-SA 4.0.
11. Map of the Karun River drainage basin, Iran. Made using public domain Natural Earth and USGS data. Wikipedia, CC BY-SA, 4.0.
12. Helmand River. Kmusser, Wikimedia Commons, CC BY-SA 3.0.
13. Map of African rivers. www.mapsopensource.com © ISS Today.
14. Total male/female body mass. © 2024 Lecturio GmbH.
15. A set of helminths: roundworm, ascaris, pinworms, bovine tapeworm, pork tapeworm, whipworm, liver fluke. Infographics Vector illustration on isolated background. iStock, credit: Timoninalryna.

16. Lymphatic filariasis. © DNA India, www.dnaindia.com.
17. Human population growth. Estimated/United Nations.
18. Thermohaline circulation. Map by Robert Simmon, adapted from the IPCC 2001 and Rahmstorf 2002.

PLATES

1. The water cycle. Source: freepik.com.
2. The structure of the earth. Source: freepik.com.
3. Water-stressed states around the world. Source: World Resources Institute, wri.org/aqueduct.
4. Provincial map of Spain. Source: Wikimedia Commons, Emilio Gómez Fernández and Javi C. S.
5. The Kingdom of Morocco. Source: *The World Factbook 2021*, Washington DC: Central Intelligence Agency.
6. A detailed illustration of the water-stress situation in Africa. Source: GRID-Arendal resources library, grida.no/publications/471.
7. Maritime chokepoints. Source: *American Journal of Transportation*.
8. Water pump on Broad Street, Soho, London. Source: Wikimedia Commons, Jamzze.
9. Mary Mallon (Typhoid Mary) in hospital. Source: Wikimedia Commons.
10. Winston Churchill, Franklin D. Roosevelt and Josef Stalin at the Yalta Conference, February 1945. Source: Wikimedia Commons, US Army Signal Corps.
11. A piece of intestine, blocked by worms and surgically removed from a three-year-old boy at Red Cross War Memorial Children's Hospital. Source: Allen Jefthas, South African Medical Research Council.
12. The schistosomiasis life cycle. Source: Wikimedia Commons, CDC.
13. How thirsty is our food? Source: Statista, statista.com/chart/9483/how-thirsty-is-our-food/.
14. Clevedon Pier. © Clevedon Pier & Heritage Trust Ltd.
15. A beach cleaner holds a handful of freshwater snails pulled from the banks of Lake Victoria in Uganda. Source: RTI International/Katie G. Nelson via Flickr.

INTRODUCTION

When I told friends that I intended to write a book about water the reaction generally ranged from gently quizzical to frankly sceptical. When I then said I believe that water will write the history of the twenty-first century, with profound implications for both humanity and the planet, it is fair to say eyebrows were raised.

However, as I explained to them that only 3 per cent of the world's total water is the freshwater we need to survive and that most of it is locked away in polar ice and glaciers they began to really listen. When I pointed out that this finite resource must serve 7.8 billion people today, compared to only 1.6 billion at the beginning of the twentieth century, and that competition for it could trigger global conflicts they started to pay attention. By the time we started to discuss how global climate change could affect the distribution of water, change the pattern of global disease and influence the potential for mass migration, they began to think that it was not such a crazy idea after all.

Water, and its properties, affects everything in our natural environment and many of the things we take for granted in the world around us. For example, while the longest day in the northern hemisphere is on 21 June, the warmest summer weather and the warmest sea temperatures are in July and August. By the same token, while the shortest day of the year is 21 December, the coldest weather is in January and February, and the coldest waters do not warm up till well after the spring equinox on 21 March. If our warmest and coldest periods were purely about the amount of energy we receive from the sun we should be at our hottest by Midsummer's Day and our coldest by Christmas. Yet, we are not. We probably never stop to think that this phenomenon is driven entirely by the physical properties of water, namely that water warms up and cools down much more slowly than land. While water has the most profound influence on human existence, we often fail to recognise its impact.

There have been three main drivers behind this project. The first is my belief that our primary, and most basic, human right must be access to clean water, not just for health but for life itself. I have spent much of my life discussing access to freedom, democracy and human rights and how we should promote and expand them. They are all important ways in which we can improve the human condition but, when it comes down to it, you can survive, however imperfectly, without them. The same cannot be said about water.

The second driver is my disgust at the way we dirty and pollute our natural environment. The trigger for this was when I first watched, with horror, pictures of the giant patches of debris, mostly plastic, floating in our oceans. Considering the life-giving nature of our great blue spaces and their ability to influence our health and well-being through currents, weather and basic foodstuffs, what we have done in recent decades is an unparalleled act of environmental vandalism, the reversal of which we should all make our business.

The third element is the sheer number of potential conflicts developing around the world which have water as a potential trigger point. All things being equal, we might be able to negotiate our way through this increasing tension between upstream and downstream nations, but the changes in global climate are making this an uphill struggle as the mismatch between population and resources becomes ever greater. If we fail to recognise the warning signs, there is a real danger that we could be sleepwalking into a nightmare.

Of course, many people have written books about water that are more subject specific, more expert and that are more substantive, but I wanted to bring a slightly different, perhaps broader, perspective to the subject, to tell the tale of water in a more complete way. This is not a book for experts as virtually everything within it is in the public domain, but I wanted to be able to join the dots in a way that doesn't always happen. It has been a source of frustration to me, not least in political life, that we tend to look at problems in isolation, in policy silos, rather than in the integrated way in which they exist in the real world.

Unless we can tell this most crucial story in a way that is comprehensible and compelling then we are unlikely to achieve the wider consensus, beyond the scientific community and subject specialists, that is crucial for wider change.

I have always had a strong interest in natural history, history and science. As an eight-year-old, I was given a book called *Discovering Science*, which awakened my interest in a whole range of subjects that eventually led me to study medicine. The net result of these interests has ultimately led to my addiction to natural history and 'how the universe works' programmes on TV. The initial wonder of how excess dust and debris turned into our beautiful world has given way to fears about the consequences of the self-destructive behaviour of our planet's most successful species.

As a doctor, I learned early on that the physiology of water balance and circulation is a key determinant of both good health and disease processes. Most people know that we are made of 50 to 60 per cent water and can last for only a short time without it. Few people, however, understand that, in evolutionary terms, we are more water efficient and some of our kidney structure is closer to aquatic mammals than any other anthropoids. This raises some interesting questions about the traditional savannah theory of human development. In other words, the story of water in human evolution is itself evolving!

As a minister in the British Foreign & Commonwealth Office, who answered for development and aid issues in the House of Commons, I was able to see for myself how much clean water and proper sewage meant to some of the world's most deprived populations. It was while launching such a project in Calcutta that I first met and became friendly with Mother Teresa. I still have one of the last letters that she wrote to me before she died.

As Defence Secretary, I was concerned by the potential for water shortages to cause conflict around the world. The dramatic growth of the human population desperately scrambling to ensure adequate water supplies for agriculture, industry and human consumption is well understood, if underestimated as a security issue, but the speed at which tensions over shared water can explode into outright conflict and the potential for military escalation is not yet fully appreciated. If countries are willing to use some of the most dangerous weaponry available if their arterial water supplies are disrupted or diverted (as Pakistan has hinted over the potential disruption of the Indus), how can we create the mechanisms for conflict resolution, including in international law? How do we deal with the growing power imbalance between upstream nations and their downstream neighbours?

As International Trade Secretary, I began to understand the

importance of water in the global trading system and how using it inappropriately could be economically and ecologically disastrous. It comes as a big surprise to most people to discover that the biggest dairy farm in the world sits in the Saudi desert and requires around 2,000 litres (L) of water to produce 1 litre of milk or that the amount of water required for the irrigation of global cotton crops alone is equivalent to twice the total annual water footprint for the entire United Kingdom.

This book is set out in four parts. The first establishes the context for the rest of the book. It looks at how we evolved from water, how earth became the blue planet and how water is distributed. Understanding that the water cycle most of us learned about in school means that we live in a closed system where water is neither created nor destroyed, brings into sharp focus the impact that an exploding population can have on fixed supply. If it seems a bit dry at times (excuse the pun) it is key to making sense of the later sections, especially those dealing with climate change. It also lays out new findings relating to the relationship between our own journey from primates in Africa to humans today. Why are we so different to other primates, how did we come to have much greater water efficiency and what did our interactions with other species, such as the Neanderthals, mean for our later development?

In Part 2 I look at the politics of water and the potential for future conflict.

It is predicted that by 2030 around 47 per cent of the world's population will be living in areas of high water stress and there will be growing tensions between upstream nations, who control water supplies near their source, and downstream nations, who depend on plentiful and predictable supplies.

Tibet is home to the greatest store of freshwater outside the polar regions and the rivers arising from it supply water for drinking, agriculture and industry to over 40 per cent of the world's population. As its glaciers start to shrink due to global warming, it does not take a genius to see where conflict for the world's most basic resource might lead. China now controls the source of all Tibet's major rivers with an ever-expanding dam-building programme to bolster its dominance. Those who believe that the Chinese obsession with Tibet is about identity, culture or the Dalai Lama should think again – this is largely about control of a single commodity: freshwater. Across the world, tensions are building. In the tinderbox of the Middle East, disputes are longstanding – between Syria, Turkey and Iraq over the Tigris–Euphrates basin, between Jordan, Lebanon and Israel over the Jordan River and between Iran and Afghanistan over the Helmand River. In Africa, the nations of Egypt, Sudan and Ethiopia compete for the control and use of the waters of the Nile while further south, disputes over the Great Lakes simmer. Will we be able to create mechanisms to resolve growing tensions or will burgeoning populations with their increasing demands for evermore water push us to breaking point?

It also considers the chokepoints in the great global maritime trade routes and their vulnerabilities, a subject that has suddenly thrust itself back onto the international political agenda with the potential restrictions at the outlet of the Red Sea, due to the actions of the Iranian-backed Houthi rebels.

The third part of the book examines how water is essential for good health, but how it can also transmit diseases that remain a tremendous scourge on many parts of the world. Water imbalance can produce problems ranging from the common hangover to life-threatening pulmonary and cerebral oedema (water congestion of our lungs or brain) at high altitudes, while waterborne diseases claim the lives of around 3.5 million people each year, with over 2 million children being lost. What can we learn from our past victories over these afflictions in the developed world and what risks are we running for their re-emergence? More importantly, how can we better co-ordinate our international response to afflictions like the neglected tropical diseases that affect some of the poorest people in our world and have a disproportionate impact on women and children? When it comes to the development of the sanitation most of us take for granted, how can the lessons of our own history be applied to the developing world, how much will it cost to give everyone access to clean water and decent sanitation and who should pay?

The fourth and final section looks at the future of water. The water we use in our daily lives cannot simply be accounted for by how much we drink, how much we use for cleaning and bathing or our regular daily activities. We need a better understanding of how the food we eat and the clothes we wear can have a massive impact on how water is used in different parts of the world and how the choices we make as consumers can make a profound difference, for better or worse, to people thousands of miles away. Before we consider the impact of climate change itself on the earth's water systems, what are the likely impacts on human health and disease, and how will this affect global issues such as migration? Finally, of course, comes the subject of climate change itself. Whatever anyone's belief about its origin, the evidence of global warming is all around us and will have profound implications, even if we arrest its current acceleration. We need to find solutions to the human and environmental impacts of the process and the potential impact on the global climate and living conditions, while simultaneously trying to control the exacerbations caused by an ever-increasing human population.

In the end, water turns out to be a much more complex and fascinating subject than it might first appear and many of my friends have concluded that it is, actually, an interesting subject to write about.

This book is not intended as a science textbook, but to tell the story of water as our story, the story of the human race and, indeed, our entire world. I hope it will provoke greater interest in the debate around the single most important issue of our time while we have time left.

PART 1

WATER, THE WORLD AND US

CHAPTER 1

WATER AND HUMAN EVOLUTION

Knowing how our planet got here, where water came from and our own place in earth's story is key to understanding both the magnificence and fragility of nature. Hopefully, it will also give us a sense of humility about humanity's place in the broader sweep of time and remind us of our duty to protect, rather than destroy, our wonderful inheritance.

TIME

To understand our relationship with the natural world around us, we need to have a feel for the concept of time. It is not an easy task to grasp the vast numbers involved as we move from millions to billions. To put things in context, the Big Bang, the beginning of the universe, was 13.8 billion years ago. The earth was formed much later, some 4.5 billion years ago, and through a range of processes, gradually developed into the blue planet that we have today. Against this enormous period of time, modern humans have inhabited the planet for a mere 200,000 years or so. How do we put this vast difference in time in a comprehensible context? We can leave the breadth of time between the formation of the universe and the formation of earth to one side for now and concentrate on the time between the birth of our planet and today.

My favourite analogy is that of the Empire State Building in New York City. It has 1,860 steps up to the 102nd floor. Let's assume that all the steps right up to the 102nd floor represent earth time with the formation of the planet at the very bottom. The earliest cellular life may have emerged around 3.5 billion years ago (although that timing remains somewhat controversial), that is, by the time you have climbed 402 steps to the 22nd floor. The first simple animals do not emerge until you have climbed 1,610 stairs to the 88th floor. If you keep climbing, you will find that fish emerge on the 90th floor, insects on the 93rd and reptiles on the 95th. We are now only seven floors from the top. The dinosaurs ruled the world for 135 million years from around 230 million years BC – that means that they own the territory between the 97th and 100th floors. The small mammals that were to survive the dinosaurs' extinction first appeared on the 97th floor, but the primates, from whom we are descended, didn't arrive until the 101st floor.

And we humans? You have to get to the top step of the top floor before you find us. In fact, if the history of the earth is represented by 102 floors and 1,860 steps, you need to reach the top 10th of the last step before we turn up. New kids on the block doesn't do us justice – we are merely a blink of an eye in earth time.

EARLY LIFE

How did this phenomenal journey from dust to humanity take place?

At first, all developing life on earth, including our ancient ancestors, lived in the sea. For about half of earth's lifetime there was no oxygen in the atmosphere until, around 2.7 billion years ago, a group of microbes called cyanobacteria evolved. They changed the course of our planet's history by generating energy from sunlight and producing bubbles of oxygen, which would eventually create the atmosphere we take for granted today.

As the atmospheric conditions changed, around 430 million years ago, plants began to colonise the planet's bare landscape and produce new food and resources that would fuel evolution and development. Fish evolved from ancient vertebrates in the sea and, 30 million years later, some of them crawled out of the water to begin the evolutionary process that would result in today's biodiversity, including our own species.

It was not an easy ride for these early lifeforms away from the comfort and protection of their watery world as they had two huge problems to confront. The most immediate was how to stop the water in their bodies from evaporating directly into the air and the other was how to ensure that their eggs would survive in a harsh, dry environment. The first problem was solved by remaining in damp and moist habitats as much as possible, which reduced water loss and led to the gradual evolution of waterproof skin. The second problem was more difficult. While amphibians, which were now also developing, were able to lay eggs in the water, the land-based species had to develop eggs that could survive in the dry if they were to reproduce and survive. The solution was that, eventually, eggs became wrapped in several layers to create what is known as an amniote egg, the same model that today sees human embryos develop within amniotic fluid contained in the amniotic sac. It was this great evolutionary development, allowing an enclosed watery environment, that enabled the emergence of the entirely land-dwelling species that would eventually diversify into today's reptiles, birds and mammals.

HUMANS ARRIVE

Having conquered the challenges of moving from a wet world to a dry one, the stage was set for the evolutionary process that would result, hundreds of millions of years later, in the development of the most successful animals of all: humans.

For the last 150 years or so there has been a consensus around what has been referred to as the 'savannah' scenario of human evolution.

This suggests that our earliest ancestors emerged around 2 million years ago, living in water-rich rainforest or moist woodland. At first, we were knuckle walkers, like today's gorillas and chimpanzees, but eventually we stood up and walked on two feet. This gave us great advantages in hunting for food across the East African savannah and enabled us to see further and travel faster. Our brain increased in size and our pelvis shortened. These anatomical differences brought their own consequences. The shorter pelvis, for example, explains why birth is more difficult in modern humans than in other primates. As our brains got bigger, but birth canals got smaller (because of standing upright), it meant that there had to be a shorter gestation period in humans, which is why we give birth to relatively immature offspring who are much more dependent on their parents than other species and are unable to walk much before twelve months of age. In today's medicine, along with its greater fear of litigation, this combination of larger heads and smaller birth canals has contributed to the huge increase in births by Caesarean section.

Generation by generation, the march of evolution went on and we continued to develop out on the grasslands and on the edge of the forests until eventually, some 70,000 years ago, we walked out of Africa and began to spread across the rest of our world. As we did so, it was not without cost for those we encountered on our travels, a pattern we were destined to repeat throughout our history.

Eventually, we would wipe out whole other groups, such as the Neanderthals, though not before they, in turn, left an indelible mark on us. Such was the preponderance of mating between our ancestors and their hairy neighbours that their genes mingled with ours. The lasting impact of this sexual imperative to reproduce is that for many of today's humans, other species, like the Neanderthals, have made a permanent contribution to our own genome. Those who have taken the now readily available home DNA tests may have already discovered this, much to their surprise. Of the 7,462 DNA variants I had tested, 256 traced back to the Neanderthals. This accounts for under 2 per cent of my DNA, although it was pointed out that I have more Neanderthal DNA than 75 per cent of the others who have been tested!

For the next 60,000 years, our population slowly increased as the Stone Age progressed. Then, around 10,000 years ago, following the end of the last Ice Age, a seismic change occurred which set the scene for the population explosion which has continued to the present day. That change was a fundamental realignment of our relationship with the world around us, namely, the mastery of irrigation and the development of agriculture. The rest, as they say, is history.

OUT OF WATER?

Nowadays, the 'savannah hypothesis' of evolution is not unanimously held and new data is resulting in revisions to that narrative. Back in 1960, the marine biologist Alister Hardy proposed what has come to be known as the aquatic ape hypothesis. It suggested that some of our human ancestors took a divergent evolutionary pathway, adapting to more watery habitats and hunting around the seashore for shellfish and other foodstuffs. Hardy's theory was not particularly well received by the scientific community at the time and has generally tended to be something of a fringe belief for the sixty-plus years since its publication. Recently, however, there has been an increasing interest in the relationship between water and primate development. It is an interest I share though I don't believe that the aquatic ape hypothesis itself is the answer, as there is insufficient evidence to support its claims.

Research published by Herman Pontzer, Associate Professor of Evolutionary Anthropology at Duke University, and his colleagues, has demonstrated that one of the features distinguishing humans from chimpanzees and other apes is our water efficiency. The human body uses 30–50 per cent less water per day than our primate relatives. This suggests that something happened during our evolution to reduce the amount of water that we need to remain healthy.

Certainly, a change in our ability to conserve water may have enabled our ancestors to travel further from watering holes and rivers and allowed them to search for food more effectively. Our thirst response may also have changed, behaviourally and physiologically, resulting in a lower craving for water per calorie. This is consistent with the fact that the water to calorie ratio of human breast milk is 25 per cent less than that of any of the great apes. In other words, human babies require less water for the same amount of energy and nutrition. It has also been postulated that when we started to develop more prominent noses, around 1.6 million years ago, we were able to conserve water by condensation and reabsorption in a way that flat-nosed chimpanzees and gorillas cannot do.

Professor Clive Finlayson, the eminent palaeontologist and director of the Gibraltar National Museum, suggested in his book

The Improbable Primate that natural selection has rewarded those best able to locate key resources like water in areas of scarcity. He argues that this was the birth of the 'rain chasers' – humans who were suited to long-distance walking and running. The increased mobility that ensued, as well as the large-scale consumption of meat by our ancestors (unrivalled among other primates), gave us a game-changing evolutionary advantage. Finlayson also argues that drying periods in the earth's climate over the past few million years were the driving force not only for humans to emerge but also for the extinction of the Neanderthals around 30,000 years ago.

When I met with Professor Finlayson in Gibraltar, he had just returned from his latest excavation in Gorham's Cave, part of the cave complex that is a UNESCO World Heritage Site and considered to be one of the last known habitations of the Neanderthals in Europe. We spent some time discussing the Neanderthals' disappearance and its implications for our own genetics and evolution.

OUR NEANDERTHAL COUSINS

The Neanderthals' disappearance is viewed by some as a true extinction. Others contend that Neanderthals did not become extinct, but were assimilated into the modern human gene pool. I think that this explanation stretches credibility a bit too much. However, it is important to grasp that the percentage of Neanderthal genes in each individual may not represent exactly the same genes. Therefore, it has been suggested that 20 per cent, and perhaps even more, of the Neanderthal genome still exists in the human population today. It is suggested that interbreeding probably took place in the Near East during an early part of the human journey out of Africa but before their arrival in Europe.

This fits with the view expressed by Professor Finlayson and others, that habitat degradation and fragmentation occurred in the Neanderthal territory long before the arrival of modern humans. According to this view, we would have arrived in areas previously occupied by Neanderthals after the latter were already extinct and the two species would never have met in Europe. Certainly, there is growing evidence to suggest that climatic instability during the millennia leading up to the Last Glacial Maximum (around 20,000 years ago) may have been a driving force in the Neanderthal extinction. Modelling of climatic stress based on new data has found that stress peaks at around 65,000 and 30,000 years ago, the second instance appearing to be more prolonged and severe than the first, and possibly related to the Neanderthal extinction.

So, was it human competition or changes to climatic conditions that caused the Neanderthals to disappear?

In medicine, we say that 'common things occur commonly' so it might be that a combination of factors – environmental deterioration as well as competition for resources – were responsible.

To add to the complexity of the argument, we know that these were not the only potential neighbours for early humans. It has been proven that Neanderthals and Homo sapiens deviated from each other around 600,000 years ago and continued to evolve in very different parts of the world. Fossils of Neanderthals have been found across Europe and Asia, in areas as remote as southern Siberia. Four-hundred thousand years ago, however, the Neanderthal line itself split, with the emergence of the Denisovan line. The Denisovans ranged from Siberia to Southeast Asia and may have persisted until as recently as 30,000 years ago, based on the genetic legacy of living Southeast Asians. Genetic sequencing from both modern humans and Neanderthals tells us that our species interbred. This might have been the result of consensual mutual courtship, or it could have resulted from something altogether less friendly – we will never know. The genetic data, however, suggest that while the two species first encountered each other when Homo sapiens made their early excursions out of Africa, the Neanderthal genes that we possess today come from much larger migrations that modern humans undertook around 60,000 years ago.

Whether or not interbreeding was successful appears to depend on several factors. Recent studies suggest that the modern human Y chromosome completely replaced the Neanderthal Y chromosome when male Homo sapiens started to mate with female Neanderthals. Since the Y chromosome is passed from fathers to sons, when male Homo sapiens mated with female Neanderthals, future generations of Neanderthals inherited the Homo sapiens version of a Y chromosome. Thus, the flow of genetic material went in one direction only. That makes it likely that while Neanderthal fathers may have produced 'human' children with human mothers, human fathers did not produce 'Neanderthal' children with Neanderthal mothers.

All this raises further questions. If modern DNA evidence is correct and not only were Denisovans surprisingly diverse, but they may have been the last humans other than Homo sapiens on earth, will we find further evidence of interbreeding in the future? It may well be that multiple lines of our ancient ancestors bred with us and that what we think of as distinct Homo sapiens may be much more hybrid than we imagined.

A WATERY PATH TO EVOLUTION?

Gorham's Cave is located on the southeastern face of the Rock of Gibraltar. Although it is only a few feet away from the Mediterranean today, when it was first inhabited, at least 125,000 years ago, lower sea levels would have meant that it was nearly three miles from the coast. It is likely, therefore, that much of the archaeological evidence relating to the Neanderthals, their lifestyles and settlements, now lies beneath the sea. Their relationship with water has been of increasing interest to scientists in recent years and provides useful clues to the story of our own evolution.

From the work of Professor Finlayson and others, we now know that Neanderthals could probably swim and dive. Evidence based on more than 170 handmade shell tools found in an Italian cave suggests that they may have also swum underwater to retrieve live clams to shape into tools. The type of clams in the area suggest that our Neanderthal cousins could retrieve shells from water that was at least 13 feet deep. Recent discoveries made during the summer of 2023 at a newly named cave in Gibraltar – the Neanderthals' Grotto – show the extensive exploitation of marine molluscs, especially limpets, by Neanderthals.

Other fossil evidence also points to the Neanderthals' aquatic lifestyle. Bony growths found in their ears are a feature they share with modern-day humans who spend time in cold and wet surroundings, leading to the condition's name of 'surfer's ear'. These growths, known properly as external auditory canal exostoses, suggest that aquatic foraging was a big part of the Neanderthal lifestyle. This appears to support the theory that at least some primates adapted to life in landscapes other than the savannah, including living near water.

This anatomical clue was recently studied by palaeontologist Erik Trinkaus and his colleagues at Washington University in St Louis, Missouri when they looked for these bony growths in the remains of seventy-seven ancient humans from the later Pleistocene period, which lasted until the end of the last Ice Age, around 11,700 years ago.

The researchers found that around half of the twenty-three

Neanderthals they studied showed signs of these growths. This was more than twice the prevalence they discovered in the other groups of ancient humans they had studied, suggesting that Neanderthals had, indeed, been water foragers.

All this was a great vindication for Peter Rhys-Evans, a leading ear, nose and throat (ENT) surgeon in London and one of the UK's foremost head and neck cancer surgeons. In a paper on 'surfer's ear' in 1992 he had predicted that if these exostoses were found in skulls from this period, it would be firm fossil evidence for the fact that our ancestors frequently swam and dived. He postulated that a phase, which saw us wading through water, could be linked to unique human characteristics such as erect posture, loss of body hair and the distribution of our body fat deposits. Perhaps most interestingly, he suggested that such a developmental pathway is responsible for the fact that we have a completely different heat regulation system to other primates and that the shape and structure of our kidneys is much more like those of aquatic mammals than those of our ape cousins. Is it possible that a marine diet rich in fish oils and proteins also fuelled the brain development that resulted in our intellectual advance and dominance as humans? It doesn't take much stretching of the imagination to believe that if our ancient ancestors did indeed walk out of the African savannah to populate different parts of the world, then many of them would have ended up on the coastlines of a world of which two-thirds is covered in water.

It seems certain that future research will help to unravel exactly how close our relationship to water has been in shaping our species and, therefore, our planet and its future.

But before we turn to current and future human interaction with our watery world, we need to explain how it became wet in the first place and what the origins of 'the blue planet' are.

CHAPTER 2

OUR WATERY WORLD

BLUE

At first, after the Big Bang, there was just dust and gas. Then, around 4.5 billion years ago, a small and distant part of this giant cloud began to collapse in on itself. The centre developed into a star, the one we know today as our sun. In time, the rest of the cloud flattened into a disc from which the planets, their moons and the asteroids in our solar system were formed. All of us – planets, moons and living beings – must, therefore, be made from the same basic building blocks. We are all, as they say, stardust.

On one of these hot, rocky planets something remarkable happened. Not too close to the sun to be scorching hot and not too far away to be freezing, the earth formed in what scientists call the habitable zone and what popular culture refers to as 'the Goldilocks zone'. The result was the creation of a unique and beautiful blue planet. The second man to walk on the moon, Buzz Aldrin, recalled how awestruck he was when 'I could see our shining blue planet earth, poised in the darkness of space.'

We are unique among the rocky planets of our solar system in having liquid water on our surface. But why exactly is our planet blue when water itself appears colourless? The reason is the way that sunlight is selectively scattered as it travels through our various watery layers, beginning with the water vapour in our atmosphere. Like all molecules, water molecules will preferentially absorb different wavelengths of light. The most easily absorbed are infrared, ultraviolet and red light, meaning that the deeper the water, the bluer it appears. That is why the shallow waters around, for example, the Aegean islands in the Mediterranean will appear azure while the vast Pacific Ocean is dark blue. Since our planet is mostly covered in water, and most of that is the ocean, we appear blue from the distance of space.

THE WATER CYCLE

A great deal of this book revolves around one basic process – the water cycle. Probably the most important point the reader must recognise is that the water on earth exists in a closed system which means little, if any, enters or leaves. It is constantly recycled so that the water we drink today may have fallen as rain that once dribbled down the back of dinosaurs or that the snow on which we enjoy skiing may have once been part of a tropical sea. The cycle really has no beginning but is a constant movement from and between the different water states – vapour, liquid or solid. It is driven by nature's great engine, the power of our sun, which heats the water in our oceans and throws water vapour up into the atmosphere where it mixes with water that is transpired from plants and evaporated from the soil.

As this all cools, it forms clouds whose particles may grow and collide to form precipitation which can fall as tropical downpours, temperate drizzle or mountain and polar snow. While some of this snow may thaw in spring to feed many of our rivers and provide much of the world's drinking water and irrigation, other snow may fall onto glaciers or the polar ice caps and be trapped there for many thousands of years. Since it covers the largest part of the world's surface, most rain falls back into the ocean, while some falls onto land becoming the surface run-off which shapes our valleys and canyons before it eventually finds its way back to the sea. Other water may accumulate in freshwater lakes while some seeps into the ground via infiltration, thus creating huge underwater aquifers capable of storing water for hundreds or thousands of years. As we will see later, the pace at which the water cycle operates can have a profound impact on our weather, a key element in the effects of climate change.

FROM DRY TO WET

But how did the earth become as watery as it is today? It is a mystery that has consumed us for centuries, and one that science is only now beginning to unravel. Since everything on our planet is composed of elements from the solar nebula, one way or another, the water that we have today must have originated from this mass of dust and gas. There have been long, and sometimes passionate, debates about whether water formed along with the earth or whether the planet began completely dry and was later bombarded with water, for example in the form of ice, from elsewhere. Some believe it is a combination of the two processes.

One of the ways in which we can start our detective work is to determine how long water has existed on earth, given that we know that the planet itself is around 4.56 billion years old.

A GEM OF A STORY

Gemstones might seem like an unusual place to begin this story, but science has a way of finding answers in the most unusual places. Zircon, which has been used as a gemstone by humans for thousands of years, contains some very interesting clues. Its chemical name is zirconium silicate and it is common in the earth's crust. While zircon crystals are usually small they can also grow to several centimetres. For hundreds of years, gem-grade zircon has been found in material deposited by rivers in places like Sri Lanka, Cambodia and Vietnam, the primary sources for the best gem-quality stones. More recently, deposits have been found as far apart as Nigeria, Madagascar and Australia.

The stones come in a wide range of colours, including blue, yellow, red and purple. The variety of their tradenames reflects the huge natural variability in their presentation. Hyacinth or jacinth stones are transparent and reddish-brown, starlite are greenish blue, sparklite are colourless and stremlite blue. Depending on the quality of the stone, prices can vary from around $50 to approximately $400 per carat and zircon is probably the only natural gemstone that comes anywhere near imitating a diamond.

It is originally formed by the crystallisation of magma, the hot liquid and semi-liquid rock that exists under the surface of the earth, or in metamorphic rocks which are formed by the transition of one type of rock to another due to very high temperatures and pressures inside the earth.

Zircons are highly durable and resistant to chemical attack, so they are among nature's great survivors. More importantly, zircon contains the radioactive uranium-238 isotope which allows us to determine its age with the precision of plus or minus a million years (which may sound like a big margin, but is extremely accurate when we are measuring it out of several billions).

In the early 1990s, a few grains of zircon were found in a sandstone in Western Australia dating back 4.2 to 4.3 billion years. Knowing the age of the zircon and studying the ratio of oxygen isotopes in samples, scientists now believe that zircon interacted with cold water on the earth's surface about 4.3 million years ago.

This suggests that liquid water and an atmosphere must have existed soon after the formation of the earth itself, much earlier than preliminary estimates had suggested and causing us to rethink previous suppositions about the state of the early earth.

CARRIERS FROM SPACE

So, if water existed on the planet relatively soon after its formation, how did it get here? It has been suggested that around 4.5 billion years ago, when the earth was around 60 to 90 per cent of its current size, the planet was bombarded by what are known as icy planetesimals and then grew by their accumulation to its current size, eventually producing surface water. (Planetesimals began as grains of ice and dust in the material orbiting the early sun which gradually clumped together and became the building blocks of the rocky planets.)

The two main candidates suspected of providing water to the earth through such a system of 'crash and deposit' are firstly, asteroids, which lie between the orbits of Mars and Jupiter and secondly, comets from the cold outer reaches of the solar system. By looking at the chemical fingerprint of different water isotopes on earth and comparing them with those of different icy bodies in the solar system, we can continue to follow the clues and build a picture of where our water may have come from.

While comets trap their water as ice, asteroids lock the components of water, oxygen and hydrogen inside minerals. Some of the most magnificent features of the night sky are key constituents of the story of how the earth was formed and its potential source of water. The brightest asteroid visible from earth, sometimes even with the naked eye, is Vesta, which constitutes an estimated 9 per cent of the mass of the entire asteroid belt. Discovered in 1807, it is the only known remaining rocky protoplanet. Much of Vesta's southern hemisphere is covered by craters created when huge amounts of material were ejected following enormous collisions one to two billion years ago. Some of this cosmic debris fell to earth as meteorites and repeated studies have shown that they have a similar distribution of water isotopes as those that exist on our planet. Of course, this is not cast-iron proof that Vesta was the source of our water, but in recent years there has been a growing consensus that asteroids are most likely to have been the primary source of the water that makes up our oceans today.

The space probe Rosetta (named after the Egyptian stone which enabled the deciphering of hieroglyphics) was built by the European Space Agency and launched on 2 March 2004. Ten years later, on 12 November 2014, Rosetta carried a lander module named Philae (named after an obelisk written in Greek and Egyptian) which performed the first successful landing on a comet in a stunning display of how far space technology had come. What Philae discovered was a difference in the chemical footprint of the isotopes (compared to that of the asteroids), suggesting that only around 10 per cent of the earth's water was likely to originate from comets (although in 2018, during the passage of another comet, analysis suggested that this number might be somewhat larger).

So, whatever the exact numbers, it seems certain that at least some water was brought to earth by asteroids which contained its building blocks: hydrogen and oxygen in mineral form. This raises another intriguing question. If asteroids such as Vesta were the foundations of terrestrial planets like our own, and contain the building blocks of water in mineral form, how much water might be locked inside the earth itself?

It is astonishing to realise that, while most of the planet is covered by the oceans, they only make up an estimated 0.023 per cent of the total planetary mass – a tiny amount. Even adding the additional water that exists in ice, lakes, rivers and groundwater, we still only reach a minuscule proportion of that total mass. It is now thought that a significant amount of water must exist in the earth's crust, its mantle and its core, and while estimates vary quite widely it is thought that around three times the mass of the oceans could be stored in the earth's mantle, while the core might contain four to five oceans' worth of hydrogen.

All of this is a somewhat complicated way of leading us to the concept of a second water cycle – the deep-water cycle which involves the exchange of water between the surface and the earth's mantle.

THE DEEP-WATER CYCLE

In cross-section, our planet consists of the outermost layer, the crust, which sits on top of the mantle and which, in turn, surrounds the molten outer core and solid inner core.

The earth's crust, the cool, hardened place upon which all life lives, ranges in depth from around 3 to 44 miles (5 to 70 kilometres), yet although it covers the whole of the earth's surface, including the continental landmasses and the ocean floor, it makes up only 1 per cent of the entire volume of the planet. Everything alive in our world sits on this thinnest of coverings.

In contrast, the mantle makes up around 84 per cent of the earth's volume. It extends to a depth of over 1,800 miles (more than the distance from New York to San Antonio, Texas) and temperatures here can reach over 4,000°C.

The outer core is around 1,430 miles (2,300 kilometres) thick, much denser than the mantle crust and is thought to consist of around 80 per cent iron. The inner core, which has an estimated temperature of around 5,400°C is also composed mainly of iron but is thought to also contain a substantial number of heavy elements such as gold and platinum.

How does this help us to understand the deep-water cycle?

Most people will have heard about plate tectonics, the movements and rearrangements of the earth's outer shell (the lithosphere) which help to explain the building processes of mountains, volcanoes and earthquakes. Our ability to reconstruct these movements gives us an insight into the evolution of the earth's surface and an ability to reconstruct the shape of previous continents and oceans. Tectonic plates are around 62 miles (100 km) thick and consist of two types of material: oceanic crust and continental crust, and while their activity began between 3.3 and 3.5 billion years ago, the earthquakes and tsunamis that show the restlessness of the deep, especially in the Pacific Ring of Fire, are a testament to their continuing activity today.

When tectonic plates collide, in a process known as subduction, the heavier plate dives below the other and then sinks into the mantle. In this way, oceanic crust, upper-mantle rock and sediments are carried deep into the earth, along with the seawater trapped in the minerals present in all these materials. The net result is that water is recycled to the planet's interior and may then play an important role in geological processes such as the formation of magma and the lubrication of earthquake-producing fault zones.

WHEN PLATES AND WATER MOVE

The effects on the surface of the ocean can be dramatic. On 11 March 2011, in Japan, the fourth most powerful earthquake ever recorded struck, generating tsunami waves that reached almost 130 feet high in some places. It is thought that this terrifying phenomenon was created because the quake happened in a subduction zone where the tectonic plate lying under the Pacific Ocean had been trying to slide under the continental plate that holds up Japan and other landmasses. Having been stuck against each other for hundreds of years and having built up enormous pressure, the plates finally moved, causing an awesome destructive effect on the surface.

As ocean water runs down into the earth's crust and upper mantle in a subduction zone, it can become trapped and, with the huge pressures and temperatures that exist there, can be forced into a non-liquid form as hydrous minerals (wet rocks) which lock the water into the geological plate. This in turn continues to edge deeper and deeper into the earth's mantle, taking the water with it and creating an enormous recycling from the surface to the interior of the earth.

Recent studies of the Mariana subduction zone in the Pacific have suggested that over four times more water is being carried to depths below 100 km than we previously thought. This created a new riddle as a long-term net influx of water to the deep interior of the planet is inconsistent with our knowledge of geological behaviour. Changes in the global sea level over the past 3 to 4 billion years have only consisted of a few hundred metres compared to the average ocean depth of around 4 km. This suggests that, over time, fluxes of water in and out of the mantle are roughly balanced with water returning to the surface from volcanic activity and ocean hotspots. Since the amount of water being carried down by subduction is much more than we previously thought, then it seems logical that we also need to increase our previous assessments of the amount of water being expelled by volcanic activity if balance is to be maintained.

This recycling of water from the surface to the interior of the planet and back again is part of the same wider closed system. Just as with the surface-water cycle, the earth operates as a closed system where water is neither gained nor lost, but where a glorious equilibrium is established between the atmosphere, the surface and the interior of the planet itself that makes our blue world possible. Since it is neither gained nor lost, how it is distributed and used, and by how many, are the key elements that will shape our future.

CHAPTER 3

ALL THE WATER IN THE WORLD

FRESH AND SALTY

Not only is water the only substance on our planet that can be present in all three states – as a solid (ice or snow), liquid (water) or gas (water vapour) but, as far as we know, earth is the only planet where water is present in all these states. Although the total amount of water remains the same, the distribution between its three different states can vary enormously, as it has throughout earth's 4.5 billion-year history.

Today, around 97.2 per cent of all our water lies in the oceans, 2.15 per cent is in glaciers and other ice, 0.61 per cent exists as groundwater, 0.009 per cent as freshwater lakes, 0.005 per cent in soil moisture, 0.001 per cent in the atmosphere and a mere 0.0001 per cent in our rivers. Consequently, and by a huge margin, the greatest amount of our water is saltwater in the seas and oceans, with an average salinity of approximately thirty-five parts per thousand. In other words, 3.5 per cent of the weight of seawater comes from the dissolved salts (over 90 per cent of which is sodium chloride), which is roughly equivalent to 34 grams of salts in 1 kilogram of seawater.

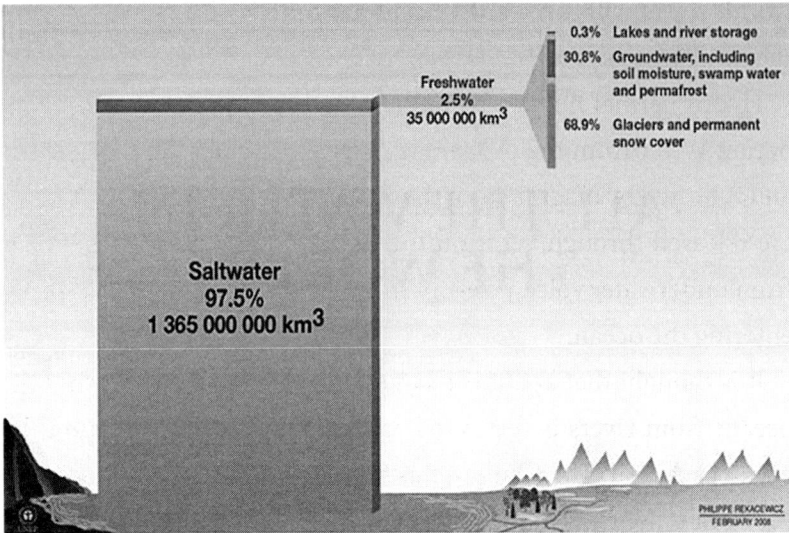

0.3% Lakes and river storage

30.8% Groundwater, including soil moisture, swamp water and permafrost

Freshwater
2.5%
35 000 000 km^3

68.9% Glaciers and permanent snow cover

Saltwater
97.5%
1 365 000 000 km^3

PHILIPPE REKACEWICZ
FEBRUARY 2008

From the point of view of human, and other animal, survival, it is a sobering thought that only 3 per cent of the earth's water is freshwater and that with much of it locked away in glaciers, ice and the atmosphere, there is only a staggeringly small amount – 0.3 per cent to 0.5 per cent – available for our use. Add to this equation the rapid increase in the human population, which currently sits around 7.8 billion, and it is easy to see why competition for water resources will be a major issue in the future, especially when we consider that it took over 2 million years to reach a human population of 1 billion but only 200 more years to reach 7 billion.

When it comes to the saltiness of the seas, how did the salt get there, how do we measure it and what does it signify?

Salt in the ocean comes from two main sources – runoff from the land and openings in the seafloor. In the beginning, the seas were probably not very salty at all, but as the rainwater that falls on land is slightly acidic, it erodes and breaks up rocks, transporting their minerals to the ocean. It is estimated that, around the whole

34

world, rivers will carry about 4 billion tonnes of dissolved salts to the oceans each year.

Hydrothermal fluids are another route by which salts reach the ocean. Water from the ocean itself seeps into cracks in the seafloor and is heated by magma from the earth's centre while salt-rich fluids are released through vents on the bottom. The release of minerals from underwater volcanic eruptions may also result in salts directly entering the ocean.

It is thought that about the same tonnage of salt that reaches the oceans from rivers is deposited as sediment on the bottom of the ocean each year. The result of this balanced input and output is that the oceans are no longer getting saltier.

Perhaps surprisingly to many, the salinity of the oceans and seas is not uniform but varies naturally in different parts of the world. Some oceans are saltier than others and even within the same oceans, some parts of the surface portions are saltier and some less salty than others. The Atlantic Ocean is the saltiest overall, although it is slightly less salty near the equator and at the poles. Why is this? The answer is that the tropics receive the highest rainfall on a regular basis and the large amount of resultant freshwater that lands there flows into the oceans, thus reducing the salinity of the surface water. In the polar regions, as might be imagined, the level of the surface water salinity is lower due to the melting of snow and ice in summer. The saltiest waters of all are in the Red Sea and the Persian Gulf, where salinity levels are close to 40 per cent because of the very high rates of evaporation and very little freshwater flowing into the seas because of their arid climates. Over the past three decades the amount of snow falling over, and into, the Arctic and Antarctic oceans has increased, providing a greater input of freshwater with the consequent reduction in the saltiness of the water. The change in density that this can produce can affect the global circulation of the oceans, and thus weather patterns, something that we will explore later.

We all know how adding salt to water reduces the temperature at which it freezes – this is the reason behind putting salt on our roads and pathways in winter. It is the same in nature, with water that is less salty freezing at slightly higher temperatures and water that is more salty freezing at slightly lower temperatures, an effect which can be clearly seen in the polar regions.

Yet there is one effect of the interaction between water and salt that might have surprised the main character in Samuel Taylor Coleridge's classic poem 'The Rime of the Ancient Mariner' when he complains of 'Water, water, every where, Nor any drop to drink', usually quoted as 'not a drop to drink'. Surprisingly, although the oceans contain a large amount of salt, sea ice contains very little. Since ice is unable to successfully incorporate salt into its crystalline structure, sea ice only contains around 10 per cent of the salt that is present in the seas. The sea that was everywhere may not have provided a drop to drink, but sea ice is actually drinkable and potentially lifesaving. It is a pity that the Ancient Mariner didn't know to save some of the ice from the Antarctic ice shelves as a perfectly usable source of water for later in his voyage.

THE DISTRIBUTION OF WATER

Today, more than 70 per cent of the planet is covered by the estimated 321,000,000 cubic miles of water that is represented by the oceans. Despite our understandable obsession with the terrestrial world, it is thought that between 50 and 80 per cent of all life on earth is found below the surface of the oceans. I still find it astonishing that in an era where we have started to explore the vast realms of space, more than 80 per cent of our ocean is still unmapped and unexplored.

Of the oceans, the Pacific is by far the biggest with 48.3 per cent of the total water on the earth's surface, the Atlantic is next with 22.4 per cent, then the Indian with 19 per cent, the Southern Ocean with 5.2 per cent and the Arctic with 1.35 per cent.

As the global climate has picked up, the resultant warming of ocean waters has produced two main consequences – rising sea levels and more frequent and more powerful storms. The increase in atmospheric carbon dioxide has resulted in increasing acidity and the increased flow of freshwater from melting glaciers may have the ability to alter the currents that affect and drive our weather patterns. While it is difficult to establish clear models that can predict exactly how these changes might manifest themselves, they represent uncomfortable, if less extreme, echoes of some of the more severe events in the history of the earth and we will look at them in more detail in later chapters.

The 2.15 per cent of our water that is locked in the planet's glaciers are mainly located in Greenland and Antarctica, although Tibet is often referred to as 'the third pole' due to the large amount of ice there. Glaciers, which can be up to two miles thick, cover around 10 per cent of the world's landmass and play a major role in climate regulation due to how they reflect the sun's light and their potential to release more water, through seasonal melting, into the oceans. Such is the amount of water locked away by this process, it has been calculated that if all the glaciers in the world were to melt simultaneously then we would see sea levels rise by around 260 feet.

GROUNDWATER
Of the available freshwater on earth, the vast majority is groundwater.

This includes soil moisture, permafrost, swamp water and the water contained in the world's aquifers. Aquifers are bodies of water-saturated rock or sediment within which water can move around freely. There are two types: confined and unconfined.

In confined aquifers there is always a layer (impermeable material such as rock or clay) that lies above and below, separating the water from the surface of the land. This leads to the water being under pressure so that if the aquifer is penetrated by the digging of a well, the water will rise above the surface and can be utilised for agriculture or human consumption.

In an unconfined aquifer (also known as a water table aquifer) the water is not pressurised but at atmospheric pressure. As a result, it can rise and fall and, because aquifers of this type tend to be closer to the surface, they are more likely to be affected by both drought and pollution which can dramatically diminish their use.

Aquifers need to be both porous and permeable, and they include a range of rock types such as sandstone (the most common consolidated rock type), conglomerate and unconsolidated sand and gravel.

Groundwater is not static and can travel through fractures in the rock or through areas that are weathered, i.e., broken down by physical, chemical or biological processes. Since rainwater is slightly acidic it can react with the limestone and cause it to dissolve, creating underground cavities and cavern systems.

The voids in the rock, created as limestone goes into solution, can weaken the structure of the land and cause collapses at the surface. These collapses are the phenomenon known as sinkholes, an item we are increasingly seeing reported in the media. Sinkholes are often a direct conduit to the groundwater and, as well as being areas which can bring human, social and economic damage are, in terms of water, areas where contamination can easily infiltrate the aquifers.

Water in the aquifers can be extremely old and even filled with what is known as 'fossil groundwater', which is defined by the United Nations Educational, Scientific and Cultural Organization (UNESCO) as 'water that infiltrated usually millennia ago and often under climatic conditions different from the present, and that has been stored underground since that time'. Confined aquifers may contain water that has been present for millennia and the fossil water that is found in some of the world's driest regions today may have been laid down in periods when they encountered much wetter and more humid periods in geological history. Some of these aquifers, in these arid climates, are being depleted of this historic resource at a rapid rate and they will not be refilled, resulting in potentially catastrophic groundwater depletion. Others are more fortunate to lie under land that receives much higher levels of rainfall, which allows them to be replenished over time.

Even where there is water to replenish them, aquifers are at risk from human pollution. Around the world, the underground storage is being poisoned by petroleum products leaking from underground storage tanks, by nitrates from the overuse of chemical fertilisers and farming, the excessive use of pesticides, and fluids leaching from landfill and waste sites.

Nearly 60 per cent of China's underground water is polluted (even according to state media) which underlines just how severe the country's environmental problems really are. According to India's National Institution for Transforming India (NITI), almost 70 per cent of all the country's freshwater, whether in aquifers or on the surface, is now contaminated. Yet, this kind of problem is not confined to poorer developing nations. In the United States, for example, the Memphis Sand aquifer in Tennessee sits below one of the country's most polluted coal ash landfills where the nearby groundwater contains levels of arsenic over 350 times that deemed to be safe.

THE GRAND AQUIFERS

The Ogallala or High Plains Aquifer sits under 450,000 km² of eight states, including New Mexico, Texas, Oklahoma, Kansas, Colorado, Wyoming, Nebraska, and South Dakota in the United States of America. It is one of the largest freshwater deposits in the world and it is thought that much of this water dates to the most recent Ice Age, or even before.

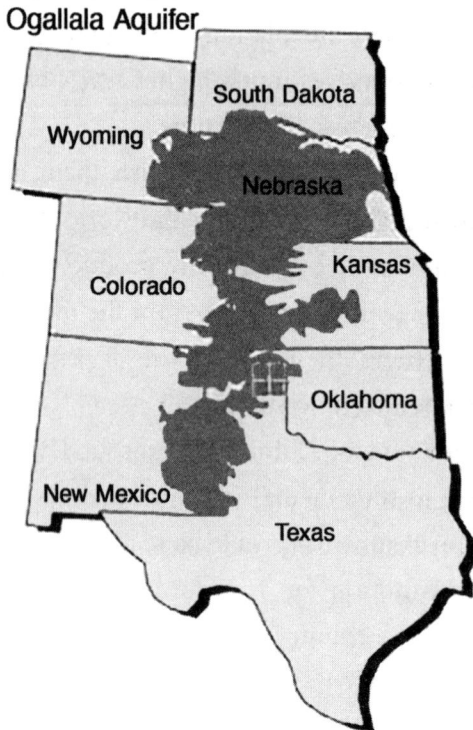

Ogallala Aquifer

Being a relatively shallow water table aquifer, the water depth ranges from around 400 feet (120 m) in parts of the north to as little as 100 feet (30 m) in the south.

It is somewhat ironic that over a quarter of the irrigated land in the entire United States lies above the Ogallala Aquifer, which today produces almost a third of all the groundwater used for irrigation in the country. Ironic, because many of the territories currently supplied by this water were part of the great Dust Bowl, the drought-stricken part of the country afflicted by severe dust storms in the 1930s. Economic depression coincided with extended drought, substandard agricultural practices, unusually high temperatures and wind erosion to lead to choking, sometimes fatal, dust storms across this same region from Texas to Nebraska. Eventually, improved conservation efforts and sustainable farming practices combined with significant amounts of rain towards the end of the 1930s to bring about the end of the Dust Bowl. However, it was only after the Second World War and the arrival of centre-pivot irrigation that this wasteland was transformed into one of the world's most agriculturally productive regions, providing the United States with one of its greatest strategic assets. Today, despite this agricultural and strategic importance, the aquifer faces the dual risks of pollution and over-extraction, with the volume of the aquifer estimated to have been reduced by almost 10 per cent since the 1950s. It is estimated that at current rates and patterns, it will take over 6,000 years to replenish the aquifer through rainfall if it becomes depleted, a prospect that puts the hardship of even the Dust Bowl years into perspective.

As groundwater becomes more of a source for drinking water, the problem of voids being created in the rock, along with sinkholes and land subsidence could well increase. The vulnerability of aquifers in this respect is clear. Once they are polluted, for example by contamination with chemicals or petroleum, it is difficult, if not impossible, to clean up.

There is also the problem of saltwater intrusion (present in coastal regions like Florida) where over-pumping of groundwater draws the denser saltwater up into the aquifer, contaminating it again.

The Guarani Aquifer is located beneath the surface of four countries: Argentina, Brazil, Paraguay and Uruguay.

At the other end of the scale, at least in terms of replenishment, the world's second-largest known aquifer is the Guarani Aquifer, which sits beneath the surface of Argentina, Brazil, Uruguay and Paraguay in South America, covering 460,000 mi² (1,200,000 km²).

Located in one of the wetter parts of the world, it is estimated that the aquifer is recharged at the rate of about 39.9 mi³ (166 km³) per year by rain and snow. One estimate suggests that this water supply is so vast that it alone could provide drinking water to the world for 200 years or even longer. It does not take a great leap of imagination to see how politically valuable (or contentious) this could become and it is likely that, as with many shared water facilities around the world, it could easily become a focal point for political tensions. As with other natural reservoirs, the biggest risk faced by this enormous natural gift is pollution by human activity.

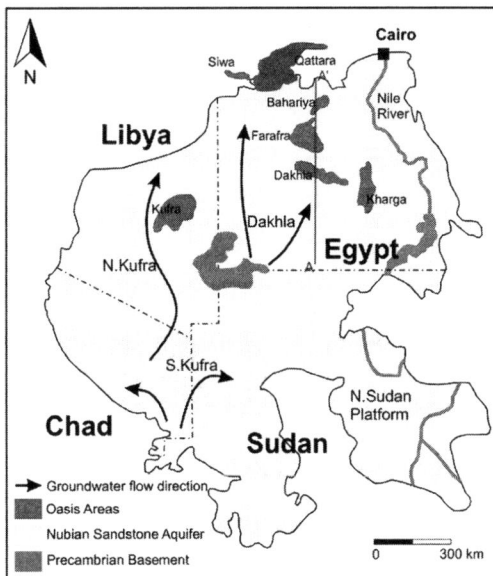

The Nubian Sandstone Aquifer System sits under Sudan, Chad, Libya and Egypt in north-eastern Africa covering around 772,204 mi² (2,000,000 km²) having been deposited between 49,000 and 20,000 years ago, and it depends on the location of the many interconnected sandstone aquifers which make up its whole.

Egypt is relatively fortunate in that it can obtain most of the water it uses from the Nile, fed by the rains in the mountains of the Ethiopian highlands, as it has throughout its long and rich history. There is, however, a limit to the amount of water that can be taken out of the river under international agreements. For this reason, and because of the growing population outside the Nile River Valley, groundwater is increasingly being used as the main freshwater resource in some parts of the country, with obvious consequences for the aquifer.

In southern Libya, water from the part of the aquifer that lies under the country is pumped across the desert to cities in the north, such as Tripoli and Benghazi on the Mediterranean coast. This fossil water was discovered in the 1950s as Libya sought to find oil reserves, and work on the huge project to transfer water across the entire country began in earnest in 1984. The Great Man-Made River (GMMR), as the project is known, provides 70 per cent of all freshwater now in use in Libya, transferring the water by pipe to a distance of some 994 miles (1,600 km).

Yet again, this precious water store which was built up during the last Ice Age is not being replenished. It has been estimated that, based on current trends, the aquifer could run out of water in as little as sixty to one hundred years, but that it could last as long as a millennium if measures are put in place to limit retrieval rates. The system also shows the vulnerability that can occur due to lack of maintenance or disruption, such as in times of domestic strife. During Libya's civil war between 2014 and 2020, neglect of the infrastructure saw over 100 of the wells on the western system dismantled, while in 2020 itself, armed militia seized a station controlling the flow of water to Tripoli and left 2 million people without water because of their actions. It is a warning about how fragile even the most technological systems can be and our dependence on physical and political stability to ensure continuous supply.

SURFACE WATER

The final type of water to consider in the distribution system is surface water, which, as well as the oceans, includes rivers, streams, lakes and reservoirs and makes up around 80 per cent of the water used daily.

Surface water is far easier to reach and so usually becomes the most common source of potable water.

Despite their huge importance to the human population, the total amount of water in our rivers is estimated at around 510 mi³ (2,120 km³), which is only about 0.49 per cent of the fresh surface water on the earth.

Rivers are generally compared to one another by what is known as their runoff, or flow of water, and it is easy to see just how uneven this is across the world. As a proportion of the world's total runoff of surface water, Asia represents 30.6 per cent, South America 27.6 per cent, North America 17.9 per cent, Oceania 14.9 per cent, sub-Saharan Africa 9.2 per cent, Europe 6.7 per cent, Australia 1.0 per cent and the Middle East and North Africa a mere 0.3 per cent. Comparing these figures to where the world's population lives makes it all too easy to see where water stress can emerge and where our problems may be building up for the future. The areas of the world which have, enviably, the greatest amount of renewable water are the Amazon and Orinoco basins with 15 per cent of global runoff and South and Southeast Asia with 18 per cent of the global runoff, including the Brahmaputra basin, the Irrawaddy basin and the Mekong basin. The ratio of water to population, however, can be dramatically different and we will look at these areas and their significance in more detail later.

The state of many of the world's major rivers is also a huge cause for concern, affecting continents in all parts of the globe. Again, India and China are high on the chart of shame. In India, the Ganges, which is the most holy river for Hindus, is the main source of water for over 2 billion people. The dumping of raw sewage, augmented by industrial chemicals, means that when this water is used for drinking and cooking (as it still is), the spread of waterborne disease and chemically induced illness is hugely increased. Many Indians believe that the river can wash away their sins but, given the damage done by the above and the attendant surface flow of plastic debris, it is currently also awash with huge risks to human life and health. Of China's many polluted rivers, the Yellow River is home to so many chemical factories and other industrial processes that the water was declared too toxic even for agricultural use. In Europe, the romantic River Danube is one of the most polluted, having the dubious distinction of containing the highest levels of antibiotics in Europe. Recently, researchers found traces of up to seven antibiotics that were greater than the level considered safe. This problem is augmented by farming pesticides, increasing river traffic, and waste seeping from Serbian factories that were bombarded in 1999.

On the other side of the world, Indonesia's Citarum River is in contention for the title of the most polluted river. Industrial waste from over 2,000 factories has led to mercury levels in the river way above what is regarded as safe. Serving an increasingly dense human population, this pollution is believed to contribute to the 50,000 deaths every year in Indonesia attributed to unclean water. In relatively water-rich South America, the Matanza in Argentina is known as 'Slaughterhouse River' because it contains the waste from the numerous slaughterhouses and tanneries located along its banks. Even the world's richest economy, the United States, makes the list of the world's most polluted rivers with the Mississippi River, one of the longest in the world, having high levels of nitrogen-based fertiliser runoff which reduces the oxygen level in the waters.

Obviously, water's physical and chemical properties can be harnessed for several other functions apart from agricultural irrigation or drinking water. Flowing freshwater can be harnessed for hydroelectric power generation, providing electricity without the unwanted side effects of greenhouse gas production, and saltwater can also be used in industries such as mining and power generation where there is enormous potential to develop the technology that will harness tidal power. I have always believed that in the long term we should seek to generate more of our energy from solar and tidal power. After all, one of the few things that we know for certain is that neither the sun nor the moon is going anywhere any time soon!

CHAPTER 4

STRESSED STATES

Today, millions of people across the world face water stress, which occurs when communities cannot fulfil their water needs, either because the supply is insufficient, or the infrastructure is inadequate. Seventeen countries (which are home to over a quarter of the world's population) face extremely high water stress. Twelve of these seventeen are in the Middle East and North Africa (MENA), a region that receives less rainfall than others and whose countries tend to have rapidly expanding and densely populated urban centres with a heavy demand for water. A report by the World Bank in April 2023 estimates that by 2050, an additional 25 billion m^3 of water a year will be required to meet the region's burgeoning needs. That is the equivalent of building sixty-five desalination plants the size of the world's largest, the Ras Al-Khair desalination plant in Saudi Arabia.

Taking a look at just a few countries which face different levels of water stress shows the varying levels of urgency, response and innovation that are being deployed in the face of a worsening crisis.

SAUDI ARABIA

The Kingdom of Saudi Arabia covers an area greater than the American states of Alaska and Texas combined but has no permanent rivers or standing lakes. Yet, anyone who has flown across the country will have noticed how, in many places, the dusty desert gives way to the vibrant green of agricultural development.

As the country's economic development gathered pace, the government began to encourage intensive modern farming methods. This was back in 2008. At that time it was estimated that there was around 120 mi³ (500 km³) of water beneath the Saudi desert, enough, as has often been quoted, to fill Lake Erie in the American Great Lakes.

The Mega Aquifer System (MAS), as it is known, lies beneath the Arabian Peninsula's deserts as well as parts of Iraq, Jordan, Oman, the United Arab Emirates and Yemen. The water dates as far back as the Pleistocene era when a much wetter climate filled up many of the world's aquifers. While parts of the southern mountains of the Arabian Peninsula see annual precipitation of rain and snow that reaches the aquifers, the rest is completely arid.

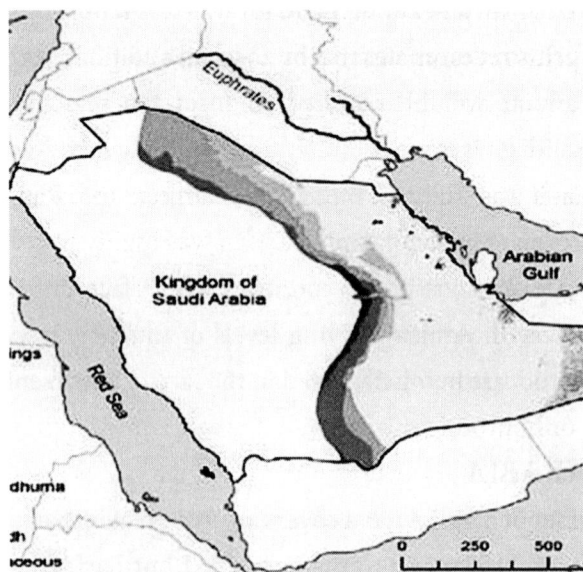

In 2015, data from the GRACE satellite programme (Gravity Recovery and Climate Experiment) showed that the Arabian Peninsula was the most stressed of the world's thirty-seven largest aquifers, with four-fifths of one of the world's oldest and largest freshwater resources emptied in a single generation.

If you drive out into the desert from the Saudi capital of Riyadh, you will find, in the Rub' al Khali desert, the site of a joint venture between the Al Faisaliah Group Al Safi Dairy Company and the French company, Danone. This is the Al Safi dairy, the largest dairy farm in the entire world. Producing around a million litres of fresh milk every day, it meets around one-third of the dairy product needs of Saudi Arabia and exports dairy products to more than 32,000 retailers in twelve countries across the Middle East region. Five-hundred-metre-long pens can each accommodate 1,500 of the 50,000 dairy cows and their offspring, who are protected from the searing desert heat by constant spraying from the so-called 'CattleCooler', which kicks in at 27°C. Since the desert temperatures in this part of the world can reach above 50°C, and are often above the 27°C level, this requires a lot of water spraying!

When I wrote about the Al Safi farm in my book *Rising Tides* in 2013, the Saudi government was considering what it should do about the fact that it was still producing its own animal feedstuff for the Al Safi cows, even though such production was highly water intensive and seriously aggravating the already depleted underwater stores. In 2015–16, Saudi Arabia, I am pleased to acknowledge, ended its domestic production which had lasted for more than thirty years and now only produces barley for human consumption. It was a correct and farsighted decision which others should seek to emulate. Consequently, Saudi wheat production, which was 141,000 MTN in 1980 and rose to a peak of 4,124,000 MTN in 1992, fell back to only 205,000 MTN in 2019. The amount of water this has saved is vast. Nonetheless, dairy cows still require a large amount of animal feed which itself requires a great deal of water for its production. In 2008, the King Abdullah Initiative for Saudi Agricultural Investment Abroad began to provide credit and political help for Saudi companies who were, and still are, purchasing foreign land and water to feed the Saudi population, including the provision of animal feed. In most of the deals, Saudi investors had generous access to water facilities and were able to export more than half of the harvest back to Saudi Arabia itself. In the case of the Al Safi owners, Al Faisaliah, they have established a large-scale agricultural business in Sudan which grows crops for both local consumption and exports. For the rest of its serial needs, Saudi Arabia has turned to more traditional markets, with Lithuania its biggest import partner, followed by Germany, Latvia, Canada and Poland.

Today, 50 per cent of Saudi Arabia's drinking water comes from desalination, 40 per cent from groundwater and 10 per cent from surface water. Even for the affluent Saudis, desalination is a highly expensive way to produce water. Saudi Arabia and the United Arab Emirates are the MENA countries making the biggest investments with $14.58 billion and $10.28 billion worth of projects respectively. There is no doubt that such means of production could be undertaken more widely across the world, but the economic and environmental costs provide their own limitations, a subject we will discuss later.

SPAIN

Not too many of the tourists who visit Spain, with its well-watered golf courses and vibrant tourist resorts, realise that this European neighbour suffers from high baseline water stress, the second-highest classification.

Back in May 2008, a shipping tanker from the south of France, with 23 million litres of drinking water on board, arrived in Barcelona to provide drinking water for its citizens, part of a plan to bring water from the Rhône every few days for several months. Thankfully, over the past decade Barcelona, has avoided a repeat of the acute crisis of 2008 but the threat of drought is ever-present. For the city itself, like many other large cities, improvements in its creaking water infrastructure are much needed, with some estimates suggesting that up to 800,000 litres of water are lost from Barcelona's ageing pipes each day.

Of course, the Iberian Peninsula is no stranger to the problem, having suffered from recurrent droughts for many decades. There are concerns, however, that the summers we are now witnessing in Spain are becoming hotter and drier because of climate change. This poses a threat not only to individual Spanish water consumers but to key sectors of the Spanish economy, such as its key tourism and agricultural industries.

The latter is particularly important for regions such as Almería, in the far southeast, where 24 per cent of the economy is agriculturally based. Rising demands for water in this area have resulted in withdrawals from the aquifers exceeding the recharge rate by up to 250 per cent. The current level of water use is around five times more than the region's annual rainfall, clearly a position that cannot be maintained indefinitely.

The degree of water stress being felt by Spain can also be measured by the levels of river pollution and the damage being done to its wetlands and national parks. The River Tagus, the longest in the Iberian Peninsula, arises in the Montes Universales in the mid-eastern part of the country and flows over 600 miles (1,000 km), close to the Portuguese capital, Lisbon, where it empties into the Atlantic

Ocean. The river is diverted and dammed in several places along its journey to provide drinking water to population centres and to help produce hydroelectric power. Around 80 per cent of the water demand on the river is for agriculture and so pollution from nitrates and phosphates used as agricultural fertilisers is widespread. Again, it is not a position that is sustainable in the long term.

Spain's wetlands have also been severely affected by demands for both river and underground water. In the southwest of Sevilla lie the swamps and marshes of the Doñana National Park which is a UNESCO World Heritage Site. Sadly, this status does not preserve the national park from the exploitation of aquifers outside its boundary or the pollution of water flowing through it, both of which have a profound impact on the supply and quality of water available. It is estimated that around 1,000 illegal boreholes have been drilled in the surrounding land for fruit cultivation, with 30 per cent of the European Union's strawberry production occurring in the area. Since strawberries are composed of 90 to 99 per cent water they are therefore a very thirsty crop. These illegal boreholes have resulted in a drastic fall in groundwater levels, with drying marshes and lagoons posing a major threat to local biodiversity.

Another badly affected Spanish area is the Tablas de Daimiel. Situated in Ciudad Real, the park is the smallest of Spain's fifteen national parks, covering an area of about 3,000 hectares (ha). Again, agricultural irrigation – arising from the need to water the vineyards in Castilla-La Mancha – has severely damaged the last example of floodplain wetlands in this driest part of the Iberian Peninsula.

In case this sounds like a story that has nothing to do with anyone but the Spanish, it is worth pointing out that tourists visiting Spain, despite the income that they generate, also greatly add to the country's water stress. Even during the Covid pandemic, when British tourists were not visiting the country, British consumers were still contributing to its water stress. A great deal of the water-intensive agricultural products find their way onto the shelves of British supermarkets, with olives, grapes, oranges, rice and numerous forms of meat understandably favoured by UK consumers. While the imports from Spain represent a relatively small element of the UK's total water footprint, the impact on Spain's water resources to produce them is severe.

It is a good example of how we might be, almost imperceptibly as consumers, worsening the position of some of the most stressed water systems on our planet without even knowing it.

MOROCCO

Opposite Spain, on the other side of the Strait of Gibraltar, only eight miles at its narrowest point, lies the Kingdom of Morocco. With the fifth largest economy in Africa, it wields influence in both Africa and the Arab world. Agriculture contributes almost 13 per cent to the country's gross domestic product (GDP), varying from modern, highly capitalised and export-orientated private farms that largely produce fruit and vegetables for export to large-scale dam irrigated areas, producing dairy, sugar, seeds and fruit and vege-tables for the domestic market. There is also rain-fed agriculture, which favours the country's wetter northwest.

The Moroccan government's 'Green Generation' strategy, govern-ing the period 2020 to 2030, sets out to develop a new agricultural middle-class representing between 350,000 and 400,000 house-holds by supporting young entrepreneurs and mobilising 1 million ha of collective lands.

The development of Morocco's agricultural sector has not, how-ever, been welcome everywhere. Since the enforcement of the association agreement between the European Union and Morocco, which opened greater trade opportunities, fruit and vegetable imports into Spain have quadrupled and doubled into the European Union as a whole. Lower wages and production costs in Morocco, combined with EU plant health and packaging regulations, have seen a downward trend in EU tomato production and foreign trade. This has led to resentment and anger among Spain's producers, in particular.

Moreover, since the re-establishment of relations with Israel, Morocco is benefiting from the latter's technology to innovate in the agricultural sector, with likely future benefits in production, as well as new market opportunities.

It is not all good news, however. Several recent droughts saw cereal production reduced by 67 per cent from 2021–22, with a consequent need to increase imports and with the Moroccan government looking to the deployment of technology as a means of providing a longer-term solution.

Morocco has approved investments in a drinking water supply costing $12 billion. It also aims to have twenty desalination plants in use by 2030, with the aim of using part of the water production to irrigate crops.

As well as increasing the water supply itself, the Green Generation strategy also aims to develop crops that require less irrigation water, such as almond and olive trees, to adapt to existing and, potentially, future droughts.

One of these new investment projects is for a desalination plant that will soon be built near the city of Dakhla in Western Sahara, which is under Moroccan administration. The city, which I visited recently, is set on a slim peninsula between the Atlantic Ocean and the Sahara Desert. Its sunny and windy climate, moderated by the

Canary Current, has made it a global centre for windsurfing and other water sports. The upwelling of the current, bringing rich nutrients from the deep waters to the surface, also sustains a strong and viable fishing industry. The future desalination plant will be able to supply 30,000,000 m³ of drinking water per year (100,000 m³ per day), and 30,000,000 m³ of this will be used for the irrigation of agricultural land, with the rest dedicated to supplying Dakhla and its growing population with drinking water. The plant will use wind electricity in this naturally windy environment to translate renewable energy into available water, using modern technology to solve an age-old problem.

A WARNING FROM THE ARAL SEA

The Aral Sea lies in Central Asia between the southern part of Kazakhstan and northern Uzbekistan. It was part of the former Soviet Union, and until the third quarter of the twentieth century, it was the fourth-largest lake in the world. Tragically, because of intense (and inappropriate) agricultural production, neglect and pollution by the Soviet authorities, the water level fell by almost 13 m between 1960 and 1987. In this period, the area covered by the Aral Sea decreased by 40 per cent – from 68,000 km² in 1960 to only 14,280 km² in 2010. Its water volume reduced from 1,093 km³ in 1962 to only 98.1 km³ in 2010, and correspondingly, salinity increased from 10 g per litre in 1962 to a staggering 130 g/L in 2010. It was one of the greatest acts of human vandalism against nature in history.

The Aral Sea is fed by the Syr Darya River from the north and the Amu Darya River from the south. In the 1950s, the Soviet regime under Nikita Khrushchev decided to promote agriculture in the area, especially the production of highly water-intensive cotton (cotton requires between 10,000 and 20,000 L of water to produce

1 kg). Water was diverted through a series of canals into the desert basin, but with such incompetence and inefficiency that huge amounts of it simply leached into the desert ground, being completely wasted. Estimates range from 20 per cent to an astonishing 75 per cent of the water disappearing in this way. As the water's salinity rose, fish inevitably died, and the fishing industry, vital to the regional economies, was destroyed. Being one-tenth of its original size, the sea has now almost split in two. The North Aral Sea, which looks like a figure of eight, lies in Kazakhstan, while the South Aral Sea, a mere strip of water in the west and a dried-out basin in the east, lies in Uzbekistan.

THE SHRINKING SEA

The changed shape of the Aral Sea since 1960

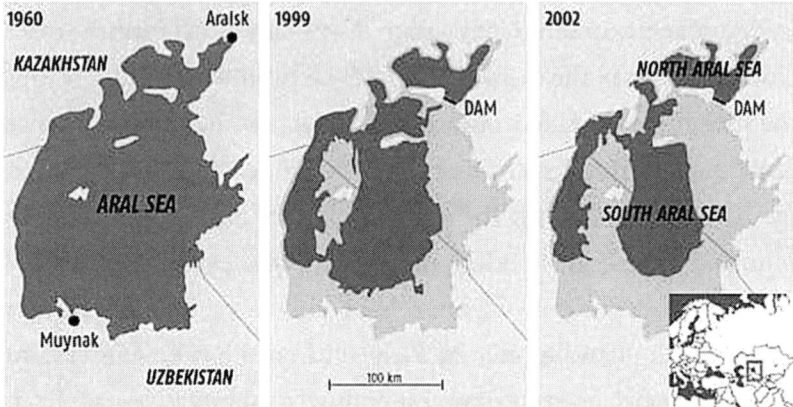

This man-made ecological disaster is estimated to have been responsible for the displacement of over one hundred thousand people. It has also had a dramatic impact on the health of the people who continue to live throughout the region. The drying out of the Aral Sea massively increased the number of dust and salt storms in the area, with the receding sea allowing the north-easterly winds to pick up sand, dust and salt and carry it straight into the lungs of those who still inhabit the region. It is estimated that more than 5 million people have had their health severely adversely affected, directly or indirectly, with increases in asthma, anaemia and tuberculosis, as well as brucellosis and typhoid. Acute respiratory disease is responsible for almost half of all child deaths in the area. Yet, despite its dreadful beginnings, this is not a story without hope. Following the collapse of communism, in the 1990s, the government of Kazakhstan decided to do something about the problem. They were helped when the World Bank contributed £66 million ($87 million) for a rescue programme that began with the construction of a 7-mile (12-km) long dyke across the narrow channel between the north and south parts of the sea. This aimed to reduce the amount of water spilling into the dry southern part and was accompanied by improvements to the Syr Darya River channels bringing water from Kazakhstan's Tian Shan mountains. The effects were more dramatic than anyone had either hoped for or anticipated, with water levels rising by 10.8 feet (3.3 m) less than a year after the project was completed in 2005. It was originally estimated that this level of improvement would take a decade or even more. The decimated fishing industry was given a new lease of life. Not only did catches increase from 1,360 tonnes in 2006 to 7,106 tonnes in 2016, but as freshwater species recovered, local fishermen were able to catch bream and catfish, attracting much higher prices and contributing to a major recovery in their income and a subsequent blossoming of the local economy. Biodiversity was boosted along the rejuvenated rivers with insect life, a necessary part of the food chain for the fish, reappearing, as well as Asian foxes and wild donkeys bringing life back to the desert landscape. This project shows what can be done.

A word of caution. The Aral Sea, whose name roughly translates to a 'Sea of Islands', referring to over 1,100 islands that had dotted its waters, may have spectacularly recovered, showing the resilience of nature even in dire circumstances – but only in part. While residents around the northern part of the sea in Kazakhstan have seen a major improvement in their environmental, social and economic conditions due to farsighted government policy and international financial support, the same is not true on the other side. The South Aral Sea in Uzbekistan remains dried out and is still producing the same combination of sand, dust and salt which poisons the lungs and damages the health and economic prospects of its population. It is a clear example of the difference that occurs when some governments take the right policy decisions and others do not. Nature can be a powerful healer and ally but sometimes it needs our help. The tale of the Aral Sea should act as a lesson that when we work with nature, water can work for and with us. When we fail to do so, it can have catastrophic consequences. The choice is ours.

PART 2

THE POTENTIAL FOR CONFLICT

CHAPTER 5

'WATER WARS' IN PERSPECTIVE

In March 2023, over 10,000 people came together from across the world, under the auspices of the United Nations, to discuss the global water crisis. In a world where we can explore the furthest reaches of our solar system and where nations are currently engaged in a new race to the south pole of the moon, it is a sobering thought that around 2 billion people on our planet do not have access to safe drinking water and that 40 per cent of the entire global population are victims of water scarcity.

As the population grows and urbanisation continues apace, it is estimated that freshwater demand will increase by more than 40 per cent from today's levels by 2051. Estimates for the global population itself are for the number to reach 9.7 billion by the year 2050 which, to put it into context, is 385 per cent of the number of people alive in 1950. This trend towards much higher populations is not evenly spread across the globe, with some countries experiencing staggering levels of growth and with some African states exceeding 2,000 per cent rates of growth in population over this 100-year period.

The UN conference looked across the range of water-related issues, emphasising the critical link between clean and available water and good health, food security and poverty reduction, as well as the role that restricted access to water can have on the growing phenomenon of forced migration, all of which are affected by the impact of climate change. Of course, we have heard lots of predictions about the potential for water to cause conflict between, and within, nations. The former vice president of the World Bank, Ismail Serageldin, is just one of the voices who have warned that wars in the twenty-first century are likely to be water-related unless significant changes occur, particularly in the quality of global governance. The fact that some of the more dire predictions have not yet come true does not diminish their significance or make them any less worrying.

It is now becoming widely accepted that such conflicts, especially at more localised levels, are seldom purely political but are impacted by demographic social and economic elements that are all, in their turn, affected by environmental and resource factors.

The climate conference ended with the United States committing up to $49 billion to support climate-resilient water and sanitation infrastructure and with the Asian Development Bank pledging $100 billion to the water sector globally by 2030. The secretary general highlighted what has come to be regarded as a fundamental human right, access to clean water and sanitation, and emphasised the need to reduce pressures on the global water system by developing new agricultural methods and making better use of recycled water, trends that have already begun in a few leading states.

WATER AND SOCIAL DEVELOPMENT

Around 14,000 years ago, the first human civilisations began to emerge along the banks of the world's great rivers, from China's Yellow River to the Mesopotamians on the Tigris–Euphrates. The rivers provided drinking water for the populations and water for agricultural cultivation, which in turn permitted and supported greater population growth. They also provided the ability to transport people and commodities, creating possibilities for human exploration and trade.

As well as utilising the rivers' surface water, people also found ways of identifying water resources underground. Roman water engineers, for example, probably the most ingenious of all, recognised that if they could see mist appearing in the early morning, low on the ground, there was an increased chance of finding natural springs or underground water resources. Others noticed that changes in soil and types of rock could indicate the presence of natural sources, while others still observed that certain plants would only grow where there was abundant underground water. All of these meant that groundwater was increasingly available to supplement the surface waters of the rivers.

The next challenge was to transport water from its source to wherever expanding and physically more distant populations needed it. Aqueducts were used in ancient Persia, India and Egypt as early as 700 BC, but it was the advent of the completely enclosed water system that ran both underground and via aqueducts, again built by the Romans, that was the most spectacularly successful. Even today, the Romans' skill in transporting water demonstrates their mastery of building and understanding of how water flows. The next great advance, the advent of sewers, allowed wastewater to be carried away from cities and prevented human waste from being merely dumped in the streets, with all the health hazards associated with that practice, especially in warmer climates. The ancient Chinese and the Romans incorporated water into their social structure by building spas and pools using water from naturally hot springs. Roman aristocrats even used water for a spot of international tourism and a tradition developed, especially for couples, to sail across the Nile, following in the historic, romantic footsteps of Julius Caesar and Cleopatra.

As seafaring peoples began to explore navigable waters, building better and stronger ships and developing navigation systems based on their observation of the stars, currents and tides, the intermingling of populations began, and international trade was born. Ancient Polynesians, who began their explorations in around 500 AD off the coast of New Guinea, crisscrossed hundreds, if not thousands, of miles of open ocean, developing new and accurate navigational systems, many of which are still used today. The mastery of water was key to increases in trade, prosperity and social development.

Yet, while the control of water was a boost to civilisations, prolonged lack of water via drought could spell disaster.

Drought deprives populations of both food and water, the essential elements for life. Under the pressures this creates, social order can eventually break down, lands can become abandoned, and even the most advanced civilisations can collapse as a consequence.

Repeated droughts, resulting in a fall of the water levels in the lakes of the Yucatán region, in what is now Mexico, saw the fall of the Mayan civilisation at the peak of their cultural development, between 700 and 900 AD. Decreased rainfall and increased evaporation due to natural climate change overwhelmed the Mayan people, who were neither prepared for the brutality of the event nor had any collective memory of how to respond to it. It has also been suggested that the deforestation practices of the Maya, aimed at increasing their crop yield in response to population growth, may have increased their own vulnerability to drought, a timely warning from history to us all.

At the same time, on the other side of the world, the Tang dynasty in China, synonymous with a golden era of literature and art, suffered severe and recurrent drought, resulting in a complete collapse by 910 AD. The collapse of two advanced civilisations almost simultaneously could hardly be a coincidence, could it? Recent interpretations of sediment records suggest that migrations in the tropical rain belt may have caused the failure of the summer monsoons around the world with catastrophic effect, again a potent warning about how even natural variations in Mother Nature can have a devastating effect on human well-being and civilisation.

WATER IN CONFLICT

Water is seldom the sole cause of conflict, but it can be one factor in several coexisting tensions. When conflict does occur, it may be due to territorial disputes where water plays a part and there is competition for strategic advantage or control over limited resources.

Such conflicts are generally categorised in three ways: where water may act as a trigger for conflict; where it may be used as a weapon to further one side; or where it is a casualty, for example, where water is poisoned or polluted.

In the first case, there may be dispute over the control of water systems or access to them, and during times of scarcity, this can lead to the eruption of violence. Conflicts in sub-Saharan Africa over grazing rights in border regions would fall into this category. In the second case, water resources or water systems are used as a tool of conflict, such as when the extremist group Islamic State sought to use water as a weapon in Syria and elsewhere. In the third case, water resources or systems may become either targets or casualties of violence, whether intentional or not. The pollution of water resources in conflicts in the Middle East comes to mind.

Conflicts can occur over both freshwater and saltwater, though the scarcity of the latter, necessary as it is for human consumption and agriculture, is a more common flashpoint. Even when there is sufficient supply for local demand, the use of water as a commercial resource including in manufacturing, recreation and tourism can provoke dispute. In saltwater disputes, fisheries have typically been the main issue. As nation states have expanded and claimed portions of the seas and oceans as their own, often protecting their own commercial fisheries, certain lucrative areas have had a history of dispute. The Bering Sea has long been a source of tension, while Britain and Iceland had a long clash over cod stocks in the North Atlantic, known as the Cod Wars.

Around the world, there are growing pressures on water due to rapid population growth, widespread environmental degradation and human-induced climate change. In the Global South in particular, there is a huge increase in consumption from an expanding middle class along with increased urbanisation in areas where poor supply brings particular difficulties. This is not only a question for the populations involved, but the richer donor countries who provide large amounts of developmental aid need to ask whether the type of help being offered is appropriate in different parts of the world where the availability of natural resources such as water may be of paramount importance.

Many experts believe that, because of climate change, not only will water supply decline in the mid-latitude regions of the world but the increasing severity of drought and flood events will alter the timing of water availability as happened to the Mayans and the Tang dynasty. Most of the population growth over the next thirty years is expected to occur in developing countries in Africa, and in

Asia, where the availability of clean freshwater is already a major issue. Uncontrolled development, combined with the destruction of natural habitats and the pollution of both surface and groundwaters is likely to increase tensions between nations, as well as lead to the increased migration of people, not least due to issues around food security. As per capita incomes rise in these developing nations there will be corresponding demands for more energy, more goods and services and more meat-intensive diets. As we will see in a later section, changes in dietary habits can cause a profound increase in the amount of water required for agriculture and food production. These factors will all lead to an increase, potentially at an accelerating rate, in water demand.

In their 2019 nature portfolio journal *Clean Water* article 'Reassessing the Projections of the World Water Development Report', Alberto Boretti and Lorenzo Rosa argued that the attitude of many in the developed world towards population growth makes it more difficult to deal with the global problem. The first obstacle, they argued, is the refusal to admit that unbounded growth is unsustainable. They wrote: 'Overpopulation arguments are portrayed as "anti-poor", "anti-developing country" and "anti-human". Population size as a fundamental driver of scarcity is dubbed as a "false notion". This denial is partly responsible for lack of good water planning.'

When combined with poverty, environmental degradation, poor quality political leadership and inherently weak institutions, this lack of planning can provide all the prerequisites, not only for conflict, but for state failure itself.

Water conflicts, of course, do not have to involve different states but can occur within a single country, such as intrastate disputes between farmers and industry over the conflicting demands for agricultural and industrial water use. In regions where water is at a premium, such as in the Middle East, where 5 per cent of the world's population has only 1 per cent of the world's freshwater, both types of dispute may exist simultaneously.

CONFLICT RISK

Countries across the world are thought to be at different levels of conflict risk. This situation reflects historical tensions and ownership claims over water resources, which are usually heightened in times of scarcity, the diversion of water by upstream powers or industrial or agricultural pollution. In the Middle East, Turkey, Syria and Iraq have long-standing tensions over the Tigris and Euphrates Rivers. Israel, Lebanon, Jordan and Syria have had historical disputes over the Jordan River basin. In Central Asia, the human-induced tragedy on the Aral Sea has resulted in conflict between Kazakhstan, Uzbekistan, Turkmenistan and Tajikistan. Perhaps the best-known and most high-profile friction is in Africa, where disputes over the Nile River have resulted in increasingly thorny relations between Egypt, Ethiopia and Sudan. All of these will be discussed in greater detail, but for anyone who has a specific interest in the history of water-related conflicts, the Water Conflict Chronology, developed by the Pacific Institute, lists disputes going back 4,500 years and cites almost 1,700 incidents of violence over water throughout the ages. The problem is not new by historical standards and many of the disputes identified by the Pacific Institute have echoed down the centuries. The bigger difficulty today is that while ancient civilisations had access to the same level of freshwater as we do, they had infinitely smaller populations to deal with. It is this growing mismatch between supply and demand that lies at the heart of many of the world's potential conflagrations today and tomorrow.

THE POPULATION EXPLOSION AND MIGRATION

The demographic changes of recent decades and their relationship to migration and conflict risk are worth a more detailed look.

Syria, which has been immersed in a bloody civil war since 2011, has seen its population increase fourfold from 5 million in 1962 to 20 million by the beginning of the conflict, with many migrating from the countryside to the cities because of the lack of freshwater for agriculture. Next-door Iraq, which has tense relations with its Syrian neighbour over the Euphrates and Tigris Rivers, has experienced a population rise from 8 million in 1960 to 40 million in 2010. In Africa, Nigeria has witnessed its population rise from 45 million in 1962 to 213 million today, while in the same period India has grown from 450 million people to over 1.4 billion. With more than half the world's population now living in urban areas, forty-five megacities each have a population of more than 10 million people, with about half of these monster urban agglomerations being in either India or China.

These population growths have been mirrored in greater levels of international migration, with people moving to places with a better quality of life and/or economic opportunity.

Europe and Asia have had the highest numbers of international migrants with around 86.7 million migrants living in Europe in 2020 and another 85.6 million living in Asia. The Latin American and Caribbean region has seen the fastest growing migrant population, while the US has the most international migrants of any single country with 50.6 million. It is followed by Germany with 15.8 million, Saudi Arabia with 13.5 million, Russia, 11.6 million, the United Kingdom, 9.4 million, France, 8.7 million and Canada, 8 million. In terms of where the migrants came from, India was the top origin country of immigrants in 2020 with 17.9 million, followed by Mexico, 11.2 million, Russia, 10.8 million, China, 10.5 million and Syria, 8.5 million. Russia, consequently, was virtually in balance between its immigration and emigration.

In 2015 a political crisis was created in Europe when over a million refugees came, in a period of months, across the borders from Syria, Afghanistan, Iraq and northern Africa. The situation has now evolved into a near-permanent crisis with economic migrants adding massively to the potential numbers from Libya, Chad, Bangladesh, Nigeria and the Gambia. It is a huge political problem with which the governments of all the destination nations continue to struggle today, a situation made worse by simplistic rhetoric. It is a basic rule in life that complex problems seldom have a simple solution. Although the issue of global migration is multifactorial, the problem has at least part of its origin in water scarcity or, at least, the mismatch between rapidly growing urban populations and the availability of water, and thereby, food supply. The economic issues around opportunity and the dangers created by protectionist attitudes in the world's wealthiest countries will be discussed later.

Yet, while the issue of migration rumbles on as a perpetual political hot potato in Europe and the United States, it is the developing countries themselves who carry the greatest burden. In 2017 it was estimated that of the 68.5 million forcibly displaced persons globally, 25.4 million were refugees, over 40 million were internally displaced and 3.1 million were asylum seekers.

MAKING A BAD SITUATION WORSE

What water issues could contribute to such an increase in human misery? There are, unsurprisingly, numerous contributory factors, many with a human cause. Landscape degradation is one of the worst culprits. In large parts of the world, overgrazing by animal stock and the removal of trees have created barren and broken land-scapes. Instead of productive pastures, the ground becomes hard and cracked and unable to retain water due to the loss of topsoil when vegetation cover is lost. Coastlines, which once saw the birth of human civilisation, are also being poisoned in many parts of the world where excessive groundwater pumping is contributing to saltwater intrusion in aquifers and ruining freshwater supplies. The Indonesian capital, Jakarta, is one such victim.

Other human activities, such as the building of dams, can also massively alter the availability of water for agricultural production. Poor governance can also be a major contributor. In my 2013 book I wrote about how disastrous water management in Syria contrib-uted to a deteriorating human and political environment which ul-timately plunged the country into civil war. Serious failures in lead-ership and policy led to attempts, not just to dramatically increase food production for the increasingly starving population, but also to produce cash crops for export. The choice of cotton, one of the thirstiest crops of all for this purpose, was a particularly foolish de-cision which had a particularly adverse effect on the availability of the already scarce water supplies. When Syria's worst-ever drought hit between 2006 and 2011, the country was particularly vulnera-ble and completely unprepared for the tragedy. The net result was over 1.5 million people moving from the countryside to the cities, adding to the already volatile social and political destabilisation. This problem was exacerbated by the huge disruption in the large global agricultural exporting countries in 2010, which had them-selves been subject to unusually severe droughts and floods. The ensuing drop in food production and rise in prices were felt, not only in Syria, but in many other north African countries, where the net result was what became known as the Arab Spring. While many western politicians chose, I believe wrongly, to interpret the regional uprisings as a spontaneous demand for liberal political rule, they were in fact a response to a commodity-driven crisis, itself a result of natural factors affecting the food producers' water supply.

Not only agriculture but industry and commerce can also be vulnerable to changes in the natural surplus or deficit of water, as we saw when extensive floods in Thailand in 2011 caused disruption to global supply chains, particularly in the automotive and light electronics industries. The 2011 tsunami in Japan saw the flooding of the Fukushima nuclear industry and triggered a global shift in policy, increasing the dependence on fossil fuels from which the world is suffering today, especially following Russia's brutal invasion of Ukraine.

We have also seen, sadly, water used as a tool of conflict, particularly in the Middle East where the Islamic State explicitly called for water to be used as a weapon, never hesitating to use the occupation and control of major dams as an important tactic in their extremist regional campaigns. They also targeted water through the disruption of water supplies, sanitation and irrigation systems which produced further displacement and migration within already traumatised populations.

THE LONG ROAD TO SOLUTIONS

The UN conference in 2023 sought to make practical progress on UN Sustainable Development Goal 6, discussed in detail later, relating to water which, if implemented, would reduce many of the underlying causes of tension that can ultimately lead to conflict. The UN goals include access to safe and affordable drinking water, access to sanitation and hygiene, improved water quality, better wastewater treatment and improved reuse rates, increased water use efficiency and improved freshwater supplies, as well as protecting and restoring water-related ecosystems.

Improved local laws and agreements can help to improve the sharing of rivers and aquifers and reduce the growing number of water conflicts that occur at a sub-national level. Importantly, strategies are also being developed in relation to international law, diplomacy and security policies, which can also be key to risk reduction. The development of fundamental principles for transboundary watershed management led to the 1997 Convention on the Law of the Non-navigational Uses of International Watercourses which establishes principles for best practice around joint basin management, data sharing and conflict resolution. While around 100 countries gathered to discuss the convention, only thirty-three had ratified it and, while this includes important players such as Iraq, Lebanon and Jordan, many of the countries who are at the centre of water basin disputes have yet to commit to the process.

Other, more specific transboundary institutions have been successfully designed, however, to promote co-operation and avoid disputes. The Indus River Commission on the 1960 Indus Water Treaty, for example, has survived, despite the continuing animosity between Pakistan and India and has shown itself to be a useful mechanism in conflict avoidance by creating a framework for consultation, the sharing of data and inspection.

Despite all the global problems, the news may be far from being all bad. The International Water Management Institute has looked at the reality of where conflicts have actually occurred compared to some predictions made in the past and believe that international boundaries ultimately and perhaps surprisingly, are more likely to become a source of collaboration and co-operation than a cause of conflict. While, they say, there have been open disputes in a handful of international basins, in around 300 shared water basins, the record has been largely positive with hundreds of treaties guiding equitable water use between nations put in place.

Why should this be? It is a fascinating feature of disputes which have been resolved that a needs-based approach instead of a rights-based approach has been adopted. In the former, the focus is on ensuring that individuals receive their minimum quantitative needs. It moves away from the position when countries view their water issues from a 'national interest' perspective. This approach guides the potential opponents away from a sterile zero-sum approach, which has typified many disputes in the past, to a much more integrative approach that allocates water and its benefits based more on human need than cold geography.

Proactive risk-reduction programmes, run by the United Nations, train water professionals in the Middle East and expand educational efforts to diplomats, civil society and legislators. By increasing awareness and knowledge of both the sorts of problems and the potential resolution mechanisms available, the UN hopes to de-escalate conflicts fairly and early in the cycle.

Other international organisations, such as the World Trade Organization, can also play a role in arbitrating water disputes when there is a commercial involvement. This role may well be expanded in the future when the role of virtual water in trade is considered more comprehensively in global trade policy, a subject that will be discussed in a later chapter.

GUIDING LEGAL PRINCIPLES

While global institutional development can assist in dealing with dispute resolution, what is not always understood is that two legal principles, that of 'prior appropriation' and that of 'riparianism' form important principles in international water law.

With prior appropriation water rights, the first person to take water from a source for beneficial use (domestic, agricultural or industrial) has the right to continue to use water for that purpose and subsequent users may only take the remaining water if they do not impinge upon the rights of the previous users. This principle is ingrained in the American legal system.

Riparian water rights allocate water, usually relating to a river, among those who have land along its path and have their origins in English common law. This legal principle exists in many jurisdictions such as Canada, Australia, New Zealand and parts of the eastern seaboard of the United States, which all originate from English common law. Under the riparian principle, landowners whose properties adjoin such a body of water have the right to make reasonable use of it as it flows through or over their properties. If there is insufficient water for all those who demand to use it, water rights are generally fixed according to the degree of frontage on the water source itself. These principles are likely to form the basis of potential settlements to some of the world's best-known, and most intractable, water disputes.

CHAPTER 6

THE HISTORIC TIGRIS AND EUPHRATES

The Fertile Crescent refers to the region that lies across today's Turkey, Iran, Iraq, Syria, Israel, Jordan and the northeastern-most part of Egypt. At its centre, the Tigris–Euphrates basin is regarded as the birthplace of western civilisation, first in its upper reaches and then spreading outwards. Remains of Homo erectus, one of our extinct ancestors and the first of our relatives to have human-like proportions have been found here. These, and our Neanderthal relatives, are thought to have passed countless millennia here, to be succeeded by anatomically modern humans. This well-watered and agriculturally rich region eventually gave rise to some of the world's first great kingdoms and empires, including the Sumerian, the Babylonian and the neo-Babylonian. These took huge leaps forward in human learning and were responsible for critical improvements in our understanding of areas such as astronomy and mathematics, as well as being the root of many of the industrial and trading practices that we take for granted today.

King Nebuchadnezzar II of the neo-Babylonian Empire, who lived from 642 to 562 BC, is remembered for his use of water in great construction projects in his capital Babylon and in his successful military campaigns. Probably best remembered for his hanging gardens of Babylon, he used water as a form of defence, building protective moats around the centre of his capital, using the Euphrates River and diverting water by the construction of canal networks.

In turn, Cyrus II of Persia, the founder of the Achaemenid Empire, invaded and conquered the city having first diverted the Euphrates and marched his troops along the dry riverbed to surprise the inhabitants at a Babylonian feast.

The Achaemenid Empire was eventually overrun by Alexander the Great and the region was subsequently fought over by the Roman Empire, the Byzantine Empire and the Sassanid Persians until the seventh century saw the Islamic conquest of the region.

One of the region's great legacies is that the ancient Sumerians, whose great civilisation blossomed between around 4100 and 1750 BC, are credited with the invention of what we understand as the measurement of time, having been the first to divide day and night into twelve-hour periods, dividing hours into sixty minutes and dividing minutes into sixty seconds.

In more modern times, following the partitioning of the Ottoman Empire (which had existed in various forms for 600 years) at the end of the First World War, the Treaty of Lausanne set out the requirement for the three great water-using states of the region – Turkey, Iraq and Syria – to reach mutual agreement before the construction of any hydraulic installation such as a dam. This also required the participation of France because of its Syrian mandate and the United Kingdom due to its mandate of Iraq, and so began the process by which multiple dams have been built on the Euphrates, sowing the seeds for many of the water-related conflicts that exist in the region today.

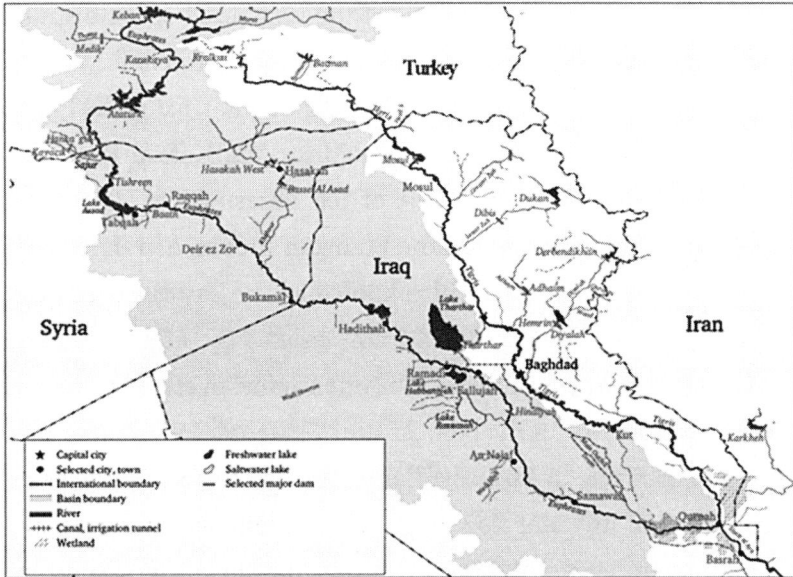

ORIGINS AND FLOWS

Understanding the origins of water flows in the Euphrates and Tigris basin, as well as the legal concepts underpinning international water agreements, is essential to understanding the complex politics of water in the region today. Both rivers originate in Turkey and eventually flow into the Shatt al-Arab basin which lies in the southern part of Iraq and is familiar to many for the role it played in the Iran–Iraq War.

The Euphrates flows from Turkey through Syria and Iraq, with Turkey contributing the vast majority (90 per cent) of the water and Syria only 10 per cent. The Tigris flows from Turkey directly into Iraq, with Turkey contributing 40 per cent of the water, Iraq, 51 per cent and Iran, 9 per cent. Syria does not contribute to the Tigris. While they may sound highly technical, these numbers are a crucial element in the complex political arguments about ownership and control of the rivers' flows and lie at the very heart of the disputes between the regional nations which can erupt at any time.

The Euphrates, western Asia's longest river, at 1,730 miles (2,780 km), originates from two main tributaries: the Karasu River and the Murat River. The headwaters of the Karasu River, also called the Western Euphrates, are in the Erzurum Province in the Anatolia region. The Murat River, also known as the Eastern Euphrates, begins in eastern Turkey near Mount Ararat, the highest peak in Turkey at 5,137 m (16,854 ft) and is often referred to as the resting place of Noah's Ark in the biblical story. These two rivers join near the town of Elazığ to form the Euphrates River. Since the river receives most of its water flow from both rainfall and melting snow, the peak volumes tend to occur during April and May. Damming of the Tigris has been important for preventing the floods that once plagued Iraq due to this sudden increase in river flow due to the rapid melting of mountain snow in April, in particular.

The increasing withdrawal of water for irrigation and the altered flows which have occurred since the first dams were constructed in the 1970s has had a profound impact on the region. The environmental effects are particularly felt in Iraq because the water used for irrigation in both Turkey and Syria flows back into the river along with agricultural chemicals such as fertiliser, lowering its suitability as drinking water downstream. The impact on the availability and purity of the water is exacerbated by the creation of reservoirs which, due to the hot climates of the region, particularly in the summer, results in high levels of evaporation which diminishes the water flow into Iraq. It is estimated that as a result of dams and reservoirs, one cubic kilometre of water is lost in Syria, two cubic kilometres in Turkey and five cubic kilometres in Iraq. Unsurprisingly, downstream in the marshes of southern Iraq, the increased salinity and decreased flow have had a massive effect on the local ecology.

The easternmost of the two great rivers, the Tigris is 1,090 miles (1,750 km) long, arising in the Taurus Mountains of eastern Turkey about 20 miles from the headwaters of the Euphrates. The river splits into several channels which feed the central marshes and the Hawizeh Marshes before joining the Euphrates near Al-Qurnah to form the Shatt al-Arab. These waterways played a major role in the first Gulf War, the US invasion of Iraq and the campaigns of Islamic State (IS) with Baghdad, the Iraqi capital, astride the Tigris and the port city of Basra on the Shatt al-Arab.

CONFLICTS

The Shatt al-Arab waterway forms part of the border between Iraq and Iran before it flows into the Gulf, and the long-running dispute over sovereignty between the two which escalated in the 1960s was one of the factors behind the outbreak of the Iran–Iraq War which lasted from 1980 to 1988. Over a million people were killed in the war, with Iran suffering the highest casualties. In recent times the two countries have concluded agreements on the delimitation and joint management of the shared waterway, but a study published by the United States Army War College concluded that 'an emboldened and expansionist Iran appears to be the only victor of the conflict'.

The Tigris–Euphrates basin can also be seen as an example of water as both a strategic asset and a weapon of war.

In the 1990s, Saddam Hussein drained the marshes of southern Iraqi Shiite Arabs who had supported the western Allies in the first Gulf War, decimating their livestock and destroying their livelihoods. His template was later taken to even more extreme levels by the Islamic State who targeted numerous strategic locations along the Euphrates. After seizing the Ramadi dam in Iraq's Anbar Province in May 2015, the water flow was reduced to below 50 per cent of its normal rate, and the southern marshes were on the edge of disaster. The targeting of water resources was part of their preconceived strategy.

Occupation of the dams allowed them to cut off water supplies to whole communities if they were not compliant with IS demands. Equally, they were able to release too much water and produce flooding to terrorise and force local communities into compliance. They were also not beyond resorting to contaminating water supplies and making them undrinkable for any population who did not bend to their extremist ideology. Drought, flood and poisoning were all waterborne weapons that they did not hesitate to unleash.

These conflicts have come at both human and environmental costs. Following Saddam Hussein's occupation of Kuwait in 1990, around 8 million barrels of crude oil found their way into the marine environment and around 3 million barrels from burning or gushing oil wells were released into the air and soil for almost ten months, resulting in badly damaged animal habitats, particularly around the oil fields themselves.

DAMS

Man-made structures altering the course of rivers in the basin are not new. In the sixth century, the 30-mile-long Nahrawan Canal, created in the sixth century, carried water from the Tigris to the Diyala River and served as the water supply for the Abassid capital, Baghdad. Water for the project was diverted by the 30-metre-high Nimrod dam, an engineering wonder of its time.

In modern times, the first water diversion structure was the Hindiya Barrage, completed in 1913 and followed in the 1950s by the Ramadi Barrage which regulated the flow of the Euphrates and discharged excess floodwater into what is now Lake Habbaniyah. Syria completed the Tabqa Dam in 1973 with Lake Assad, its reservoir, an important source of water for drinking and irrigation.

Turkey finished the construction of the Keban Dam in 1974. While this was primarily built to generate hydroelectric power for Istanbul and Ankara, it has also been important in providing water for both Syria and Iraq during periods of drought. While not designed as such, the dam has effectively become part of Turkey's GAP, the south-eastern Anatolia Project (*Güneydoğu Anadolu Projesi* in Turkish). Originally conceived and planned in the 1960s, the project was designed to produce twenty-two dams and nineteen hydropower plants, harnessing the potential of both the Tigris and the Euphrates Rivers. It remains one of the world's most comprehensive development projects, affecting around 10 per cent of both Turkey's population and surface area. As well as boosting irrigation and agricultural production in the region, the project was also designed to deal with a key political problem, namely, Turkey's problem with its Kurdish population, by reducing the income disparity that existed between the region and the rest of Turkey. The idea was essentially that the region's socio-economic development would promote market integration and reduce the influence of the strong tribal affinities that exist there. In other words, it would reduce tensions by the 'Turkification' of the Kurds.

With the UN predicting that the flow of both the Euphrates and Tigris could markedly decrease in the years ahead, resulting in a diminished quantity of water available for Iraq and Syria, the case for a well-managed water system in the basin seems to possess an overwhelming logic but the problem occurs in the difficult political relationship between Turkey and its neighbours.

A DIFFICULT HISTORY

To understand the interaction of water and politics in the region, one of the first things to grasp is that the sovereign states we think of as today's Turkey, Iraq and Syria are relatively new entities. Indeed, it is possible to argue that for over 2,000 years the Hellenistic, Turkic, Iran-Iraq and Arab empires that dominated the region imposed a substantial level of unity. This was shattered by the reconfiguration of the region after the First World War and though many of today's tensions may appear to originate from issues around water, much of this may be a proxy for competitive sovereign and nationalist interests.

Following its defeat in the First World War, the Ottoman Empire was dismembered. The Treaty of Versailles not only completely abolished the empire but required Turkey to renounce all its rights over Arab Asia and North Africa. Under the mandate system, which was effectively an internationally approved legal form of colonialism, Britain was granted Iraq and Palestine while France gained Syria and Lebanon. The result was the emergence of a new state based on Turkish national identity and, at the same time, Arab states with a sense of identity unable to govern themselves independently. Resentment at this position fostered an increased sense of Arab nationalism that would reach its peak following the Second World War and the departure of the French and British. What was missing, and is a continuing issue to this day, was any concept of a Kurdish state. The net effect of all of this was the eventual emergence of new and competitive sovereign states with accentuated national identities and a Kurdish population with a keen sense of identity but no state at all.

Coinciding with these seismic changes was a breakdown in legal commonality. The various empires who had governed in the region brought with them common systems for the arbitration of disputes over water. With the establishment of the new republican Turkish state and the post-Ottoman Arab states, neither the waters nor the people of the Tigris–Euphrates basin came under the common rule of Islamic jurisprudence any longer.

Until the 1960s the three countries sharing the Tigris–Euphrates basin: Turkey, Iraq and Syria had a reasonably quiet relationship over water management, but rising energy requirements, the increasing need for water for agriculture, growing human populations and external political factors began to change the equation and led to conflicting demands over the allocation of river water that each nation should be given. But how could this be done?

ALLOCATING WATER RIGHTS
Several elements need to be considered if there is to be a rational allocation mechanism acceptable to all. The first is to look at how much each country contributes to the water system. As stated previously, Turkey contributes 90 per cent of the water flow to the Euphrates, with Syria contributing 10 per cent. Iraq does not contribute at all. In the case of the Tigris, Turkey supplies 51 per cent and Iraq 40 per cent. Syria does not contribute at all.

Turkey argues that it produces the vast majority of the water and is therefore entitled to its share based on the 'prior appropriation' principles discussed earlier.

Iraq and Syria argue that they enjoy water rights that go back to the history of the ancient world and that since both the Tigris and

Euphrates were historically shared watercourses, today, they should both be classified as shared resources. It is a good example of a classic upstream–downstream dispute. Essentially, Syria and Iraq argue that they are entitled to calculate their own water requirements and that these figures should then be used as the basis for allocating river flows. Unsurprisingly, this is completely unacceptable to Turkey.

The next element to consider is the water requirements of each nation.

There are many ways to measure how water 'rich' or 'poor' a country may be, but combining total water availability with population size gives an idea of the social and political stress felt by individual states.

Measuring annual renewable freshwater in cubic metres per capita allows us to see developing patterns. In 1970, Iraq had 10,304 cubic metres per capita, Turkey had 5,682 and Syria 7,367. By 2001, the numbers have fallen to Iraq having 4,087, Turkey 3,029 and Syria 2,700. The projected numbers for 2025 are Iraq 2,392, Turkey 2,356 and Syria 1,701. The drop in the availability, per capita, for all three countries is deeply worrying.

It is widely accepted that water-rich countries are those which have 10,000 m^3 of water per capita annually, so all three countries are water stressed by comparison. Therefore, it is also not true that Turkey has abundant water while Arab countries do not when population sizes are taken into account.

We can see the next level of complexity in the picture when we look at how much water is being demanded from each of the river systems. Syria seeks 32 per cent of the Euphrates and 5.4 per cent of the Tigris. Iraq wants 65 per cent of the Euphrates and 92.5 per cent of the Tigris. Turkey intends to use approximately 52 per cent of the Euphrates and 14.1 per cent of the Tigris. So, the combined demands amount to 248 per cent of the Euphrates' total flow capacity and 111 per cent of the Tigris'. Clearly, these demands are a mathematical impossibility.

Out of these issues, the one that is likely to prove the thorniest is making an objective assessment about which land is most irrigable and will, therefore, produce the greatest return for 'water investment'. Classification of land in this regard is often divided into six categories, the top three of which concern efficiency and who can produce maximal yields, category four is of marginal value, the fifth requires considerable investment to produce yield and the sixth is land that is infertile even with available irrigation.

Turkey argues that all its land to be irrigated by the Euphrates is in the top three categories while only 48 per cent of potential agricultural land in Syria reaches this criteria. It would be wrong, and wasteful, they maintain, to irrigate infertile lands at the expense of fertile ones. While there is undoubtedly some technical merit in this case, it is politically impractical unless the yields are equally shared.

In the case of the Tigris, Turkey utilises virtually none of the water while Iraq utilises its entire annual capacity. When Turkey proposed to ease the water shortage of the Euphrates by transferring water from the Tigris, Iraq unsurprisingly objected.

In fact, Syria and Iraq have strongly objected to all the water installations that Turkey has either planned or completed on the basis that these would reduce the quantity of water flowing into their territories, or would at least leave them at a strategic disadvantage should Turkey decide, for whatever reason, to restrict the flow.

Turkey's response to this would be that the dams already constructed not only contribute to Turkey's irrigation and energy needs, but that they have been able to regulate water flow on rivers where the natural fluctuations have historically produced an excess of floods and drought. Moreover, many hydrological experts would agree that it makes sense to build dams further upstream. Evaporation losses are less and geographical and topographical features tend to make construction easier and more efficient.

So, how do we balance the competing arguments from Syria, and to a lesser degree Iraq, that because of their perceived antique acquired rights, the rivers must be shared according to a quota that reflects a simple mathematical formula and Turkey's assertion that the Euphrates and the Tigris must be regarded as one single transboundary watercourse system, including the man-made Tharthar Canal connecting the two rivers? This takes us back to concepts of usage rather than ownership, a view that is closer to Turkey and the wider international community than it is to the traditional arguments of Syria and Iraq.

Of course, rational proposals have to be viewed against the historical relationship between Turkey and its Arab neighbours. Not only is there the traditional enmity that arises from the relationship between the Ottoman Empire and Arab states, but the latter are acutely aware that almost all Arab countries are downstream states and achieving settlements on issues that might set precedents for them is always a forceful consideration. Even relatively modest actions by the Turks have induced disproportionate reactions from some Arab states. For example, when Turkey interrupted water flow during the impounding of the Atatürk Dam in January 1990 it was interpreted in some Arab circles as a clear act of aggression, even though Turkey had given its neighbours due notice about the necessity of restricting the flow during this critical period of construction.

One final factor to be considered is the issue of the Kurds. As already discussed, the Turks have sought to deal with the problem of Kurdish nationalism by economic means, integrating them more fully into Turkey through the GAAP programme. Turkey's problems, however, have been used as levers by some of its neighbours. The refusal to negotiate fully with Syrians on new water projects saw Syria turning to the Kurdistan Workers' Party (PKK) as a means by which it could wrest concessions from Turkey. When the civil war in Syria broke out in 2011, the Damascus government effectively pulled out of its previous arrangements with Turkey and switched sides to ally with the PKK. This action has been met by Turkey's increased tendency to consider its own interests first. Thus, the initial approach taken by the governing Justice and Development Party (AK Party) of achieving a situation where there were 'zero problems with neighbouring countries' has given way to more scepticism about Syria as a reliable partner and the emergence of new, more nationalist, postures. Syria and Iraq also tend to claim that the west has been too pro-Turkish in the disputes over water because, during the Cold War, Turkey was an active participant in NATO, while Syria and Iraq maintained a more pro-Soviet position.

So, what of the future? There is a widespread view that climate change could impact the basin particularly severely, with some estimates suggesting that temperatures could rise by up to 1.2°C by mid-century. This could result in lower rainfall levels with some experts suggesting that the runoff in the upper Tigris River basin could fall by as much as 30 per cent after 2040.

Clearly, a proper trilateral agreement needs to be reached if drought and ecological degradation are to be avoided and this can only occur if there are improved political relations and greater trust than exists today is built.

Combining the waters of the Euphrates and the Tigris would make it much easier for all three countries to advance their irrigation plans along the Euphrates but any arrangement will need to be equitable and deal with water usage on the basis of need, not ownership. It will also need to be something more than a short-term fix and include agreements that will be able to withstand not only the vagaries of regional politics and the changing climate but will also be saleable to the future generations of all three nations.

CHAPTER 7

THE HOLY RIVER JORDAN

The Jordan River has the lowest elevation of any river in the world and forms a small segment of the East African rift system that runs from southern Turkey via the Red Sea into eastern Africa itself. The rift is where the African tectonic plate is splitting into two plates, the Somali plate and the Nubian plate at the rate of about 7 millimetres (mm) per year.

As well as its huge geopolitical importance, the river has long held religious significance. The Book of Joshua tells of how, after the exodus from Egypt, Joshua led the Israelites across the river into Canaan. Joshua 3:9–10 says: 'Tell the priests who carry the ark of the covenant: "when you reach the edge of the Jordan's waters, go and stand in the river."' The significance of this resonates in the fact that it was the site chosen by John the Baptist for the baptism of Jesus. The UNESCO World Heritage Site known as 'Baptism Site "Bethany beyond the Jordan"' sits on the east bank of the river and has been a holy site since the Byzantine period and possibly before. According to Matthew's Gospel, 'When Jesus came out of the water', having been baptised in the Jordan by a reluctant John, 'the heavens opened. He saw the spirit of God coming down and resting on Jesus like a dove. A voice was heard from heaven. It said "this is my son the beloved. I am very happy with him."' The site has Roman and

Byzantine remains, remnants of churches and a monastery, as well as hermit caves and baptismal pools, confirming its long standing as one of the principal Christian sites of pilgrimage.

WHERE GEOGRAPHY AND POLITICS COLLIDE

The Jordan River basin has a total area of around 18,500 km². Forty per cent of the basin lies in Jordan, 37 per cent in Israel, 10 per cent in Syria, 9 per cent in the West Bank and 4 per cent in Lebanon. The Jordan River itself runs 223 km from its origins in the Anti-Lebanon and Mount Hermon mountain ranges to its ultimate discharge into the Dead Sea.

In the north, the Hasbani, the Banias and Dan Rivers, themselves fed by groundwater and seasonal surface runoff, form the upper Jordan River which flows south into the freshwater lake, the Sea of Galilee (also known as Lake Tiberias), which provides the largest freshwater storage capacity on the river. The Lower Jordan River receives its main flow from the Yarmouk River. This originates in Jordan, first forming the border between Jordan and Syria, then between Jordan and Israel before it joins the southbound outflow from the Sea of Galilee, forming the Lower Jordan River, on its way to discharge into the Dead Sea. A warning about the current problems of the basin can be seen in the fact that while flow rates from the Jordan River into the Sea of Galilee remain much the same as they were fifty years ago, the Lower Jordan River's discharge into the Dead Sea can be as little as 10 per cent of its previous value.

Surface water accounts for around 35 per cent of the resources in the area, groundwater aquifers account for 56 per cent, while wastewater reuse counts for 9 per cent. At the northern end, the Upper Basin has around 1,400 mm of annual precipitation while at the southern end the rate is only 100 mm.

Water quality and salinity vary enormously as you travel north to south.

Israel is the biggest user of water from the Jordan River basin, withdrawing around 600 million cubic metres (MCM) per year and is the only user of water from the Sea of Galilee. Jordan uses about 290 MCM of water, with diversion from the Yarmouk River to the King Abdullah Canal providing irrigation for crops in the Jordan Valley and for domestic use in the capital, Amman. Although Syria has no direct access to the Jordan River it has built several dams in the Yarmouk River sub-basin and uses around 450 MCM, mainly for agricultural purposes.

The climate, low precipitation and growing populations, along with the failure to achieve common agreement on water resources has made water a critical issue in a region that already has more than its share of political tensions.

If we go back to our international comparisons and consider countries to be 'water poor' if the available freshwater per capita is less than 1,000 m^3 per year, we can see the scale of the problem. While this value was 119 for Israel in the year 2000, it had fallen to only 81 by 2020. Likewise, it fell for the West Bank and Gaza in the same timescale from 278 to 169. For Jordan, the equivalent figures were 135, falling to a mere 62.

While this value was 254 for Israel in the year 2000, this has now fallen to only 138. In Jordan it has fallen from 234 to 61, in the West Bank it has fallen from 115 to 89 and in Gaza from 150 to 83 with the rapid population growth exacerbating an already adverse natural scenario in some of the world's most water-stressed nations.

In 1970, the population of Israel was 2.7 million. Today, it stands at 9.174 million. In 1970, Jordan had a population of nearly 1.55 million. Today, it is 11.33 million. While Israel's population growth was mainly due to migration from around the world, the huge increase in the Jordanian population is largely a consequence of massive regional migration, especially by Palestinians, and a high subsequent birth rate.

The politics (and water politics) of the region are entirely shaped by its history, as any visitor to the region will attest.

On 2 November 1917, while the First World War was still causing carnage across Europe, the British Foreign Secretary, Arthur Balfour, wrote to Lord Rothschild, a leading member of the British Jewish community. The letter stated that

his Majesty's government view with favour the establishment in Palestine of a national home for the Jewish people and will use their best endeavours to facilitate the achievement of this object, it being clearly understood that nothing shall be done which may prejudice the civil and religious rights of existing non-Jewish communities in Palestine.

The boundaries of Palestine were not defined and the phrase 'in Palestine', being deliberately ambiguous, was taken to mean that not all of Palestine would be covered by the 'Jewish national home'.

When the Ottoman Empire was dissolved after the First World War, Britain was awarded the mandate for Palestine by the League of Nations, which intended for the UK to administer the territory until it was 'able to stand alone'. Far from being a solution to the problem, this was merely the start of an even bigger complication for the British.

Both the Balfour Declaration and the British mandate created tension between Arabs and Jews and between Britain and both groups, satisfying the ambitions of neither.

Arab anger morphed into the Arab revolt in Palestine (1936–39) and Jewish resentment turned into the Jewish insurgency in Palestine (1944–47).

On 29 November 1947, the UN adopted a partition plan for Palestine, intending to divide it into a Jewish state, an Arab state and a *corpus separatum* under a special international regime encompassing the cities of Jerusalem and Bethlehem. On 14 May 1948, Israel declared independence and, on the following day, the civil conflict turned into what has become known as the first Arab–Israeli War. The net result of the conflict was that Israel ended up controlling not only the territory demarcated by the UN for itself, but also

60 per cent of that identified for the Arab state, including West Jerusalem. Transjordan took control of East Jerusalem and the West Bank, formally annexing it in 1949, and Egypt took control of the Gaza Strip.

Around 700,000 Palestinians fled or were expelled from the territory that became Israel and around the same number of Jews moved from the Arab streets to Israel in the three years following the war. The conflict officially ended when Israel signed armistice agreements with Egypt, Lebanon, Transjordan and Syria in 1949, relinquishing control of almost 80 per cent of the British-administered mandate, a third more than allocated by the UN partition plan.

These deep and painful divisions made it impossible for the states in the region to follow a single plan for their limited, but shared, water resources and set the tone for simmering tensions that would lead to further conflict later. Unilateral projects, utterly unco-ordinated with their neighbours, first took shape in 1951 when Jordan announced its intention to divert part of the Yarmouk River via the East Ghor Canal. Israel began the construction of its National Water Carrier programme in 1953, which opened in 1964, diverting water from the Jordan River Valley. This aroused fury among the Arab states and their response was to produce a plan to divert the headwaters of the Jordan River to Syria and Jordan. The Israelis, in turn, saw this as a threat to the viability of their independent state, as it would have diverted 35 per cent of the water that they intended to withdraw and when they attacked these projects over the next two years, the scene was set for the major conflict that has shaped the region today, the Six-Day War of 1967.

From the late 1940s to the early 1960s, the United States attempted to mediate. Eric Johnston, a special representative of President

Eisenhower, devised a plan that would see 55 per cent of available water in the basin go to Jordan, 36 per cent to Israel and 9 per cent each to Lebanon and Syria. While Israel was willing to negotiate, at that time the Arab countries did not recognise Israel and feared that any support they gave to the American plan would amount to recognition of the country as an independent state. Despite the failure to implement the agreement, it served as a general guideline that was tacitly followed during the 1960s and beyond.

Although the Arab states ultimately abandoned the diversion effort, it was the issue over access to different water that was to be the trigger for the next major conflict, the Six-Day War, fought between Israel and a coalition of Arab states, mainly Egypt, Jordan and Syria, from 5 to 10 June 1967.

The Straits of Tiran, named after the small island at its inlet and now overlooked by the Egyptian city resort of Sharm El Sheikh, are the narrow sea passages linking the Gulf of Aqaba and the Red Sea. They provide access to both Israel's only Red Sea port at Eilat and Jordan's only seaport of any kind at Aqaba.

By 1967, 90 per cent of Israel's oil (mainly from Iran) passed through the Straits and the Israeli government declared that any attempt to blockade them would be an act of war. In May 1967, following Egyptian demands, the UN peacekeeping force (UNEF) withdrew and on 22 May, Egypt, under General Nasser, despite all the warnings, blockaded the Straits.

The Israelis launched a series of pre-emptive strikes on 5 June, destroying almost all of Egypt's air assets and creating massive air superiority for themselves. At the same time, Israel attacked the Sinai Peninsula in Egypt and the Egyptians occupied the Gaza Strip. Jordan and Syria joined the conflict after several days, but such was Israel's military dominance that Egypt and Jordan agreed to a ceasefire on 8 June and Syria on 9 June. Just as in the first Arab–Israeli War, the result of the Six-Day War was that Israel controlled even more territory. By the end of the conflict, Israel also held Syria's Golan Heights and the Jordan West Bank, including East Jerusalem.

Following Israel's dramatic victory, Jewish immigration to Israel increased hugely in the following years, while anti-Semitism in European communist states mushroomed, only twenty years after the Holocaust. There was a huge displacement of Palestinians from the Occupied Territories and around 300,000 more people settled in Jordan, adding to the already stressed water demands. Israel and Egypt reached a peace agreement in the Camp David Accords of 1978 and Israel withdrew from Sinai 1982. Jordan and Egypt withdrew their claims to sovereignty over the West Bank and Gaza.

While Israel increased its use of water from the Jordan River by a third in the year following the Six-Day War, Jordan lost significant access to water at a time when its population was rapidly increasing. A key turning point came in the summer of 1992 during the Middle East peace process, when Israel agreed to reduce its water use and increase diversion of the Jordan River to the Yarmouk, giving Jordan much greater capacity to meet its water needs and paving the way for the peace agreement signed in 1994.

The political fallout from all these events is still being experienced today. The West Bank Valley is key to Israel's supply of groundwater. The largest aquifer in Israel, the western aquifer (or Yarkon-Taninim aquifer), has 80 per cent of its recharging area within the West Bank, whereas 80 per cent of its storage area is within Israel's borders. The strategic importance of the Golan Heights, in terms of the Jordan River's headwaters, was made clear to Israel in the period leading up to the Six-Day War and there is little or no chance of Israel relinquishing such a, literally vital, strategic advantage.

In 1995, as part of the Oslo peace process, the interim agreement provided more water from Israel for the Palestinians, with the condition that they refrain from drilling any new wells in the Mountain aquifer. While agreement with Jordan was reached as part of the Israel–Jordan peace treaty, the surface water of the Jordan River remained a source of dispute between Syria, Lebanon and the Palestinians.

FROM DROUGHT TO PLENTY

Rainfall in Israel is inconsistent, to say the least. It can fall below the multiyear average for several years in a row, but it can also be well above normal. This creates problems in supply due to a growing population and the increased demands of greater urbanisation and industrial development and everything is at the mercy of Mother Nature.

The early years of the twenty-first century saw a decade-long drought affecting the entire Fertile Crescent. The Sea of Galilee, fed by the Jordan River, Israel's biggest source of freshwater, fell to levels approaching the 'black line' at which salt infiltration would flood the lake with irreversible consequences. Israel reacted with huge public campaigns to save water. The amount of water allocated to agriculture and industry were reduced and prices for domestic customers were raised to restrict demand. Considerable financial resources were dedicated to finding ways to improve efficiency, perhaps the greatest of which was the development of the refined drip irrigation system, delivering water directly to the roots of plants.

Things were bad for Israel, but for Syria they were much worse. As described earlier, the serious water catastrophe was not just an act of nature but of political folly, with poor policy decisions aggravated by incompetence and neglect. Wells were dug deeper and deeper in a literal race to the (watery) bottom. As a consequence of this, and the regional drought, Syria's farmland collapsed into dust. Papers from 2015 in the proceedings of the National Academy of Sciences concluded that 'the rapidly growing urban peripheries of Syria marked by illegal settlements, overcrowding, unemployment and crime, were neglected by the Assad government and became the heart of the developing unrest'. Syria's continuing civil war, worsened by the intervention of outside actors such as Russia and Turkey, have made the continuing water crisis an almost permanent feature of the country's ongoing problems.

While Syria suffered, Israel acted, beginning with an intensive wastewater recycling programme alongside the installation of low-flow toilets and showerheads nationwide. The National Water Authority designed water treatment systems that now capture 86 per cent of the water that is used and recycled for irrigation. It's worth considering that the Israeli rate of 86 per cent, the highest in the world, is around four times more than any other country in the world. The United States' level is less than 10 per cent.

As ever, no solution is perfect, and there are now some concerns about the potential influence of contaminants that are not completely eliminated by wastewater treatment plants and which might find their way into agricultural products through irrigation.

The real game changer for Israel, however, was the move to create a major water desalination industry. Plants were constructed in Ashkelon (2005), Palmachim (2007), Hadera (2009), Sorek (2013) and Ashdod (2015). There are two chief methods of desalination: evaporation processes, which are an older and more expensive technology and really only suitable for countries where capital is very cheap; and membrane processes, a more modern technology which usually uses reverse osmosis with saltwater squeezed through membranes that only permit the passage of water, while salts are prevented from passing through. Israel's desalination plants are based on the latter technology and the country now has more water than it needs, an astonishing turnaround from the drought-driven fear of the past.

Perhaps the greatest example of this reversal can be seen in the Sea of Galilee. In December 2022, a groundbreaking project was launched to pipe desalinated water into the Sea of Galilee to maintain water levels even during dry years, while still satisfying the country's demands for freshwater. This is a far cry from the days of almost reaching the black line.

The national water carrier, Mekorot, also constructed an 8 mile (13 km) underground pipe network connecting the lake to the five desalination plants on the Mediterranean. When the project is completed, new desalination plants, along with water from wells in the north will further replenish the lake.

Desalination, however, does not come without its complications. Although the water is softer, reducing maintenance costs for both industry and private consumers, the complete removal of minerals has created some new health issues. The lack of magnesium, when desalination is the only source of drinking water, is associated with an increased risk of heart disease and a public debate is underway about whether this mineral should be added to the water. A lack of iodine in desalinated water also seems to be contributing to a high rate of serious iodine deficiency among the population, with the biggest impact being on pregnant women and school-aged children. Rising rates of foetal hypothyroidism have again opened a national debate about how this can be redressed in the population.

Israel's new water surplus opens the potential for water-sharing agreements to reduce regional and political tensions. The country sells about a hundred million cubic metres of water annually to the Palestinians, who themselves drill about 160 MCM in the West Bank and 200 MCM in Gaza, but the real beneficiary of a new water dynamic, apart from Israel itself, is likely to be Jordan.

ANOTHER WAR

On 7 October 2023, the Iranian-backed Hamas terrorist group launched an attack against Israel, murdering around 1,200 Israeli civilians and taking 240 hostages, including Israeli soldiers, men, women, children and foreign citizens. In response, Israel launched a major military operation, which is thought to have resulted in around 30,000 deaths in Gaza, mainly among civilians, in another tragic cycle of violence in the region. The territory's water supply was an early victim, with the Gazans access to water from desalination and external Israeli sources estimated to have reduced by 95 per cent after 9 October. Even before the war, Gaza had a huge water deficit. About 90 per cent of its water supply comes from the coastal aquifer basin running from Egypt through Gaza and into Israel on the eastern Mediterranean coast. The water is of a poor quality, with seawater intrusion as well as sewage and chemical infiltration which results in Gazans having to depend on small-scale desalination units, unregulated private water tankers and three Israeli pipelines which transport about 10 million m^3 of water into Gaza from Israel each year.

Cutting off Gaza's water supply actually shows some of the limitations of weaponising the resource. It is highly unlikely that turning off the supply will cause Hamas to surrender to the Israel Defense Forces any time soon but it will, and has, created misery for the people of Gaza, increasing the risk of another generation becoming radicalised and fuelling international criticism of the

Israeli government. Of course, this may be the political outcome that Hamas and its backers seek but there are other forces that remain intent on charting a long-term course for peace and who have not allowed the tragic violence that has unfolded to alter their course.

But a Note of Optimism?
What will happen after Gaza remains to be seen, but the territory will need to be reconstructed with new security and political structures put in place.

ANOTHER NOTE OF OPTIMISM

One of the most significant, and least celebrated, moves towards peace in the region came with the negotiation of the Abraham Accords. Negotiated and supported by the United States, the agreement brought about full diplomatic relations between, initially, Israel and the United Arab Emirates, then Bahrain and later Morocco. The official signing ceremony was hosted by the United States at the White House on 15 September 2020. Many believe it merited much greater international plaudits than it received at the time and that politics took precedence over outcomes.

One of the agreement's first tangible benefits to the region came at the COP27 climate summit in Egypt when what has been dubbed 'project prosperity' was signed. This comprises two entities – Prosperity Green, a solar plant to be built in Jordan, and Prosperity Blue, a desalination project to be built in Israel. Under the deal, the United Arab Emirates (UAE) would provide the capital to build a solar power plant in Jordan that would produce 600 megawatts (MW) of solar energy to be exported annually to Israel. The profits of the electricity sale will be shared between Jordan and the UAE and in return, Israel will send 200 MCM of desalinated water to Jordan. This energy for water swap, engineered and funded by the UAE is, I believe, one of the most important agreements reached in the region for decades. The political risks are not inconsiderable, particularly for Jordan, where many in the majority Palestinian population believe that their political interests may be compromised for wider considerations but, given that they live in the world's second most parched nation, they will also be major beneficiaries of this initiative.

CHAPTER 8

CHINA AND ITS NEIGHBOURS

HISTORICAL TRENDS

Water availability or scarcity per capita is dependent, as we have seen in the Middle East, on natural resources, population growth, urbanisation trends and geographical variations in supply. All of these patterns and trends can be seen in monumental scale in South and East Asia between China and her neighbour India, as well as Pakistan and nations of the Mekong basin. China's population has risen from 654 million in 1960 to 1.264 billion in 2000 and 1.425 billion in 2023.

India has experienced a similarly gargantuan growth spurt, from 445 million in 1960 to 1.059 billion in 2000 and overtook China with 1.428 billion in 2023. In the same period, Pakistan's population has grown from 45 million in 1960 to 240 million in 2023.

Demands on water have been further massively magnified by rising levels of urbanisation.

In 1949 when the People's Republic of China was founded only 10 per cent of the population were urbanised. This grew to 37 per cent by 1986 and over 50 per cent by the turn of the century. In 1911 only 10.29 per cent of India's population was urbanised, rising to 31 per cent by 2011. By 2035 it is expected to be 43 per cent, with

635 million Indians living in organised communities compared to China's 1 billion in similar circumstances and with the number of megacities growing accordingly.

Unsurprisingly, these trends have had a major impact on the availability of water. In 1961, China had 4,280 m^3 of water per capita available, which fell dramatically to only 1,993 m^3 per capita in 2020. India's figure has also fallen to 1,486 m^3 per capita today and is expected to drop to only 1,367 m^3 by 2031. Next door, the situation is even worse. Pakistan's availability per capita, which was only 1,500 m^3 in 2009, fell to a meagre 1,017 m^3 in 2021, perilously close to water distress.

Even these numbers do not, however, tell the full story as there are huge regional distribution differences within countries. For example, 80 per cent of China's water resources are in the south. The dry north depends on depleting groundwater resources while the south is supplied by surface water from rivers which are themselves increasingly stressed from both flow rates and pollution perspectives.

SAME WATER, MORE PEOPLE

One of the major sources of water in the region is Tibet, which was occupied by China in 1950 and whose rivers provide freshwater to almost 2 billion people. Considering re-usage and agricultural methods, it is estimated that 46 per cent of the world's population depend on rivers originating in Tibet, from the Indus and the Ganges to the Mekong. China, India, Bangladesh, Pakistan, Vietnam, Burma, Cambodia, Laos and Thailand all depend on rivers which originate in Tibet so what happens there has, and will have, a major geopolitical impact. Countries around the Tibetan Plateau are all experiencing the same rapid urbanisation as that occurring in China and India with millions migrating from the countryside straining the water infrastructure of cities and emerging megacities even further.

With no water sharing plan in place, just as in the Middle East, unilateral water policies are creating ever-increasing tensions. China, India, Pakistan and others have plans for hundreds of hydropower facilities along the rivers emerging from Tibet, many of which will have a huge impact on the ecosystems downstream, bringing new social and interstate frictions. Industrial projects are diverting water for their own economic ends and increasing deforestation, soil erosion, ground pollution and unpredictable weather and flood patterns all add to the risk of potential conflict coming in their wake.

The freshwater demands of China's rapidly expanding population for both drinking and agriculture have been gigantic, but if we add to that industrialisation at breakneck speed, we can get an idea of the size of the crisis facing China's water supply. Sixty per cent of China's 661 cities face seasonal water shortages and over 100 cities have severe water constraints according to the World Bank.

Eighty to 90 per cent of groundwater and 50 per cent of river water is unfit for drinking and half of the country's aquifers are too polluted to tap for industry or farming. Half of the population cannot access safe water for human consumption and two-thirds of rural populations are forced to rely on polluted water.

The dumping of industrial and human waste into the water supply is producing a health disaster that is exacerbated by poor environmental regulations, a lack of enforcement and political corruption. Villages around large industrial complexes have become known as 'cancer villages' with high levels of arsenic, fluorine and sulphates producing horrific rates of stomach, oesophageal and liver cancers.

The dumping of fertilisers is also leading to widespread

eutrophication, the process by which a body of water becomes overly enriched with nutrients and produces the growth of simple plant life such as plankton or algae. Over two-thirds of China's lakes are affected, as are 850 out of 1,200 monitored rivers. Lake water eutrophication is now one of the main factors impeding sustainable economic growth in China with damaging algae blooms turning lakes bright green and interfering with the water supply to cities. Each destructive bloom is estimated to cost billions of yuan in damage.

A similarly drastic situation has been created by the over extraction of groundwater. This has not only seen seawater intrusion and ecological destruction but has produced more and more land subsidence, with ninety-six cities affected, including the capital Beijing where, on its east side, the level of subsidence reached 10 centimetres (cm) per year between 2003 and 2011.

The reliance of China's industrial base on coal-produced electricity creates another problem, since coal-fired power plants need a lot of water for cooling. This alone accounts for around 12 per cent of the country's water consumption and is one of the reasons for the push to hydroelectric power. An additional complication is that two-thirds of China's coal plants are in the heavily water stressed northern areas. Overall, China's highly polluting industrial base, containing steelmaking and chemical and paper production, consumes 22 per cent of the country's water and is driven by the global demand for cheap consumer goods.

FEAR OF DROUGHT

This chronic shortage of water can be critically compounded by periods of drought. In 2022, China was hit by its most severe heatwave in sixty years resulting in a record-breaking drought. Parts of the Yangtze River, which provides water to 400 million people, dried up, affecting shipping and hydropower and limiting drinking supplies, with water levels in parts of the river at their lowest since 1865. This drought was in a region that depends on dams for three quarters of its electricity generation so polluting coal stations were brought back into play. In Chengdu, a city with a population almost the same size as London, power was out for up to ten hours a day with Volkswagen and Toyota both forced to temporarily close their factories. Crop damage in the affected area, in southern China's Yangtze basin, worsened the economic hit. It reminded me of a conversation I had as shadow Foreign Secretary with President Hu during a visit to the UK. The discussion had been around North Korea's nuclear ambitions, but when I asked him whether he was more afraid of the strategic threat produced by his next-door neighbour or drought in China, he and his officials laughed at what they assumed was the rhetorical nature of the question. Drought was, is and will continue to be, a major threat to China's social stability and the future of the Communist Party apparatus.

China's policy response to its water crisis is officially set out in what are known as the three red lines. The first is controlling the development and utilisation of water resources, the second is improving water use efficiency, bringing the level up to that of the world's best by 2030. The third is limiting pollution entering rivers and lakes and ensuring that compliance in monitored water zones is 95 per cent. More widely, however, the policy has been to acquire, divert and dam.

ACQUIRING MORE

The acquiring part of the policy began in 1950 when Mao Zedong invaded and annexed Tibet, boosting China's strategic position and water resources simultaneously. It increased its landmass by more than a third and suddenly made China a neighbour of India, Nepal and Bhutan. By giving China control of river flows from the Tibetan Plateau, it enabled the country to utilise Tibet's water for its own internal purposes while simultaneously becoming Asia's chief upstream water controller with a powerful lever over those nations dependent on rivers flowing from the plateau. An additional bonus for China is that Tibet is an Aladdin's cave of mineral resources, the world's largest lithium producer and the reserve of ten different metals for China's economy and military. China's mining and damming activity now threaten the ecosystems of the world's highest and largest plateau with drastic, and potentially irreversible, effects.

The arid northern regions of China have been at the centre of its plans to divert water from the Tibetan Plateau and from its own wetter southern parts. More than twenty physical water transfer projects, including the immense south to north water transfer project, have already been put in place to help water the northern region which provides more than half of China's wheat, a critical strategic reserve in a country that is unable to produce enough food for its own people.

CHINA'S RIVERS – AND DAMS

Of course, two of the great rivers that arise on the Tibetan Plateau have played a huge role in the economy, culture and history of China itself. The Yangtze River, which arises in the high Tanggula Mountains of northern Tibet, is the longest river in Asia and the third longest in the world, flowing for 6,300 km across China to the east coast near Shanghai. The Yellow River, which at 5,464 km is the second longest river in China, arises in the Bayan Har Mountains and flows across nine Chinese provinces before flowing into the Bohai Sea in Shandong Province. Mighty as these rivers are, they cannot satisfy China's growing thirst, and it has increasingly turned to one area in which it has great expertise – dams. China has long been the global leader in dam building and already has more than half of the world's 58,000 large dams. Still, many more are planned, and every time a dam is built on the great international rivers that flow from the Tibetan Plateau, China enhances its ability to use the ultimate threat, limiting the flow of transboundary water, as a devastating political weapon. China has, for example, already constructed seven dams along the Mekong River in Tibet and more than twenty further dams are planned. This means that in Cambodia, Laos, Thailand and Vietnam, 60 million people who are dependent upon the river for food and water security, have to increasingly look over their shoulder at China's potential coercive abilities.

Inside China, the Three Gorges Dam, 181 metres high and 2,335 metres long, spans the Yangtze by the town of Sandouping in Hubei Province. The dam, which cost US$22.5 billion, took seventeen years to build and required 1.3 million people to leave their homes, can generate 22,500 MW of electricity, making it the most productive hydroelectric dam in the world. It was highly controversial both in and outside China and thousands protested against the destruction of ancestral homes, the flooding of fertile farmlands and the complete destruction of communities, with protests more recently about the increased threat of landslides. The Yangtze Plain, however, had long been prone to serious flooding, including in 1931 when up to 4 million people are thought to have perished so, by reducing the downstream flood risk, as well as creating a massive source of electricity generation, the Chinese authorities regard the project as a great achievement.

Other projects are also highly controversial with China's neighbours and international conservation bodies. The plan to build thirteen hydropower dams along the main stem of the Salween River is one such scheme. The river flows over 2,800 km from its source in the Tanggula Mountains (which also give rise to the Yangtze), crossing Tibet and southwest China before flowing into Myanmar where it forms part of the border with China and empties into the Andaman Sea. A World Heritage Site, the Salween is home to around 25 per cent of the species of animals in the world, including the most diverse population of turtles. Altering the free flow of the water into a series of small channels and reservoirs could have enormously damaging ecological consequences in the extensive drainage basin, home to more than 7 million people in Myanmar.

Not all countries, however, have been hostile to China's dam building projects. Thailand buys around 3,000 MW of Chinese hydro powered electricity and a major deal has been struck with Nepal, although critics have been quick to question whether there have been sufficient environmental impact assessments carried out and whether there is competent risk management in areas where the seismic activity is well above average.

China's main problem with its neighbours, however, is one of trust. The country's unwillingness to entertain resource-sharing agreements is a major obstacle to co-operation and does not bode well for negotiating a solution should there be tension between China and any of its smaller neighbours. The understandable fear is that China's need to maintain its hydroelectric output means that they will withhold water during the dry season, making an already difficult situation even worse for downstream countries, which have good reason to be sceptical. The 1995 agreement on co-operative sustainable development of the Mekong River basin accepted that the Mekong does not belong to any single state, an attempt to deal with the sensitivities of transboundary governance. China flatly refused to join and two years later voted against the United Nations Watercourses Convention, again refusing to be part of a multinational attempt to achieve consensus on how transboundary issues should be handled. Time and again it has cited its own sovereign interests and rights as the reason for its failure to work with international partners. It is clear to any objective observer that China is making clear that it will place its own interests first, irrespective of the impact that has on its neighbours, damage to its international reputation or any concept of international law.

In any case, many will wonder what value can be placed on a Chinese signature on any international agreement following China's blatant disregard of its own obligations, particularly under the treaty signed with the United Kingdom over the status of Hong Kong. Here, despite the commitments entered into by China, there has been a brutal crackdown on political dissent and freedom, making a mockery of the one country, two systems concept.

While, understandably, a great deal of international attention is paid to China's increasingly flagrant disregard of international law in the South China Sea, perhaps, if we want to see how Beijing really operates, more attention should be paid to the other side of China, in Tibet.

TIBET: 'WHAT EVERYONE WANTS'

Tibet has been a key part of international trade since ancient times, including as part of the south-western route of the Silk Road. Tibetan salt, wool and precious stone traders would descend to Milam, the first village on the Indian side of the plateau, to sell their goods into the global markets and it is clear from Ptolemy's Ganges Delta map that the river's route was well understood from its source in Tibet to what is today's Bangladesh. Throughout history, because of its natural resources and its strategic position, it has been, to one degree or another, 'what everyone wants'.

In the nineteenth century, the whole of the southern flank of Tibet came under the control of the British Indian Army. In 1903, after pressure from Lord Curzon, the British Viceroy of India, an expeditionary force was launched to counter what was perceived as Russian influence and a threat to British India. It was, effectively, an invasion force and reached Lhasa, the capital, in August 1904. The invasion was one of the factors for the 1905 Tibetan rebellion when many foreign citizens, including missionaries, were slaughtered. It was not until 1967 that a British officer would have a permanent posting in Lhasa again.

INDEPENDENCE SNUFFED OUT

Tibet came under the control of the Qing Dynasty of China in 1720 and remained as a protectorate until 1912. After the Xinhai revolution in 1911 ended China's last imperial dynasty, it effectively became a self-governing entity, coming under Tibetan government control by 1917 and independent from the rest of the Chinese Republic. Sadly, this was not to last. Despite attempts by the Tibetans to retain their independent status, by the late 1940s the Chinese were circling. In September 1949, just before the proclamation of the People's Republic of China (PRC), it was the Chinese Communist Party's (CCP) explicit intention to incorporate Tibet, along with Taiwan and Hainan into the PRC to expand its territory and thereby improve its defensive (and economic) position. Since China's new revolutionary leader, Mao Zedong, knew that the Tibetans would not willingly hand over their independence to the CCP he made plans to invade. The British and Indian governments made vain attempts to try to intercede but were stonewalled by a Chinese three-point plan which insisted that China be responsible for Tibet's defence, trade and foreign relations and that Tibet must be part of China. Failure to agree to these points would lead to war. After several months of failed negotiations, the People's Liberation Army crossed the Jinsha River into Tibet on 6 October 1950. China's attention was now focused on what it called 'the Seventeen-Point Agreement' and hostilities were broken up during what were almost laughably known as negotiations, which were effectively a 'take it or leave it' offer from China with clear implications of what 'leave it' would mean.

Essentially, the Seventeen-Point Agreement meant that in return for accepting that it was part of China, the Tibetans would be allowed to reform at their own pace, with self-government over their internal affairs and religious freedoms. The pattern of behaviour that was to be repeated over half a century later with Hong Kong is obvious. The myth of one country, two systems was all too clearly shown for the fiction it always was in Tibet long before the Chinese decided to disregard their treaty agreement with the United Kingdom and effectively close down democracy and freedom in the Hong Kong territories. The Chinese had promised the people of Hong Kong they would enjoy the capitalist system and freedoms that were not found in other parts of China for fifty years, but in 2020, the imposition of dramatic national security laws brought an end to the pretence that China was willing to see any deviation from rigidly imposed Communist Party control. The same approach can almost certainly be expected by Taiwan, whenever the government of China believes that the international conditions will allow them to get away with it.

In Mandarin, the Tibetan region is known as Xizang or 'Western treasure house'. As mentioned earlier, it is a massive store of mineral wealth for China with deposits of iron, zinc and cadmium and the biggest source of Chinese copper production at Tibet's Yulong mine.

Yet, for all its economic value, its increased territorial size and defence utility, it is Tibet's water that remains China's primary concern, and there is no way it will relinquish it. In 1962, China fought a border war with India over a small, frozen piece of territory, known to the Indians as Arunachal Pradesh. The fact that this dispute remains unresolved to this day, with neither nation willing to yield, shows that both understand the full geostrategic importance of the Tibetan Plateau.

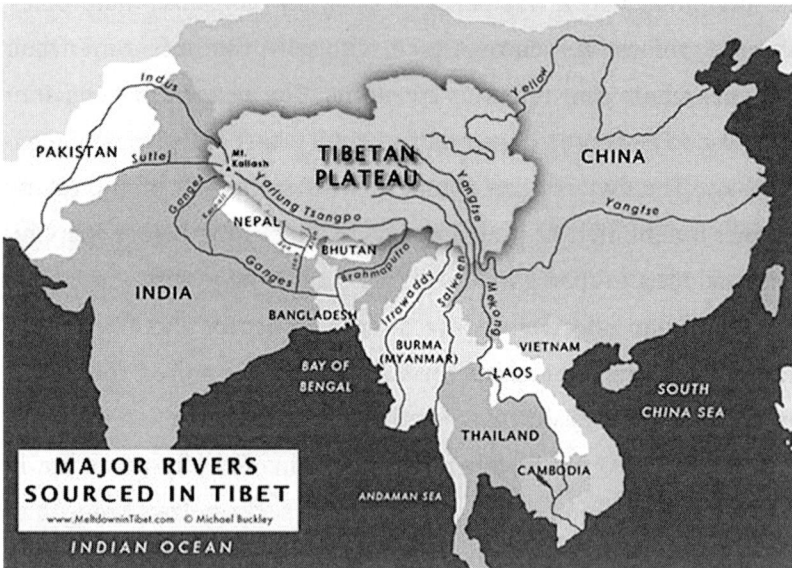

MAJOR RIVERS SOURCED IN TIBET
www.MeltdowninTibet.com © Michael Buckley

THE WATER TOWER: FLOOD INTO DROUGHT

The highest and largest plateau on earth, Tibet has a total area of some 2,500,000 km² at an average altitude of 4,900 m. It contains the

Hindu Kush Himalaya (HKH) ice sheet and is the third largest store of freshwater on the planet. After the Arctic and Antarctic, it holds the largest number of glaciers and snow, which is why it is sometimes known as the world's third pole or, alternatively, Asia's water tower. In total, the Tibetan Plateau has more than 14.5 per cent of the world's total glaciers, some 46,000, but unlike the Arctic or Antarctic, its glaciers do not flow directly into the sea but form the source, via melting snow, of ten of the world's greatest rivers. As Asia's other major source of water, the monsoon, is available for only a few months of the year, the annual melting snow is a key water source for hundreds of millions of people across South and East Asia.

Worryingly, around 247 km² of glacial ice has disappeared every year for the past seventy years. The Tibetan Plateau has seen temperature rise by 1.3°C over the past fifty years, three times the global average. This means that the plateau's glaciers are depleting faster than anywhere else on earth. In the past few years, Tibet has seen record numbers of floods and landslides, as well as substantial increases in lake volumes in different parts of the plateau, intimating further potentially life-threatening floods should the water suddenly break out and spill into lower territory.

It is a profoundly disturbing and distressing revelation that, as the rate of glacier melt increases, there will initially be more floods in the region due to increasing water flow until the glaciers melt entirely, at which time the flow will dramatically decrease, or even stop. This will have a devastating effect on the populations downstream and their abilities to maintain drinking water, agriculture and industry. Their very survival as viable communities will be brought into question, just as that of the Maya once was. There is a growing consensus that the amount of runoff water from melting glaciers will probably increase until around 2050 when it will decrease, potentially catastrophically.

In January 2019, a group of experts came together under the International Centre for Integrated Mountain Development (ICIMOD) to produce the first Hindu Kush Himalayan assessment report. This suggested that even if global warming was held at 1.5°C, a third of the glaciers would be depleted by 2100. More recently, a study published in *Nature Climate* by scientists at Penn State, the University of Texas at Austin and Tsinghua University, examined in detail how climate change is affecting the Territorial Water Storage (TWS) of the Tibetan Plateau. TWS is an aggregate measure of the total water found in any habitat. Observations and information from the Gravity Recovery and Climate Experiment (GRACE) satellite have enabled TWS to be assessed in a much more accurate way than ever before. The team found that TWS had been greatly depleted in some parts of the plateau in recent years but had increased in others, likely due to the competing effects of lake expansion, glacial retreat and the degradation of seasonally frozen ground. In particular, massive water declines in the Amu Darya and Indus River basins (which supply water to Central Asia and Afghanistan, and to northern India, Kashmir and Pakistan respectively) could have an even more disastrous impact on downstream areas and populations than previously anticipated. It is now thought that if the global temperature increases by 1.5° C, only between 37 and 49 per cent of glacier mass will remain by 2100.

It is feared that change of this scale would have a profound effect on hydrological cycles and, since the Tibetan Plateau influences atmospheric circulation (therefore weather and climate patterns) across the northern hemisphere, the impact of warming on the Tibetan Plateau could be felt much more widely than just in the immediate area itself.

While most of the focus on climate change has been due to

atmospheric changes, a recent report has suggested another reason for why the plateau is warming faster than other parts of the world. According to the World Bank, the black carbon dust that settles on the ice absorbs more heat and therefore causes it to melt faster. It is thought that fine particles of black carbon, which form through the incomplete combustion of fossil fuels and biomass, may be responsible for more than 50 per cent of the accelerating glacial and snow melt in the region, reinforcing the need for moving away from fossil fuel-powered vehicles and traditional forms of industry.

The Great Rivers that have their origins on the Tibetan Plateau are literally vital for a staggering proportion of the people alive on our planet today. It is estimated that 20 per cent of all humans are directly dependent on them for drinking water and over 45 per cent are dependent on them for this as well as agriculture and industrial use. So, the rivers of Tibetan origin are, literally, vital for China, India, Nepal, Bangladesh, Pakistan, Bhutan, Vietnam, Cambodia, Laos, Thailand and Myanmar. Some countries are more vulnerable than others, depending on how much of their water flow originates entirely beyond their borders. This measure, known as 'the dependency ratio', scores 91.4 per cent in Bangladesh and 77.7 per cent in Pakistan.

THE GREAT RIVERS

As well as the Yangtze and Yellow Rivers, already discussed, the Indus runs west from Tibet to Kashmir and then, entering Pakistan from the north, runs its entire length to Karachi and the Arabian Sea. At 3,100 km long, it is not only Pakistan's longest river but critical to agricultural and food production (as it has been for more than 4,000 years), especially in the lower Indus Valley which has very poor levels of rainfall. In *Rising Tides* I set out just how important the water supply is for Pakistan and how the country has indicated its willingness to go as far as using a tactical nuclear weapon if the river was ever to be diverted without their consent, such is the existential nature of the threat.

The Sutlej takes a different course into Pakistan and joins the Indus for its last 1,000 km. The Irrawaddy River, Myanmar's main river, has its source in the Dulong Jiang in Tibet. The Arun River has its source close to the northern slopes of Mount Everest before it runs through Nepal, and the Salween River, 3,289 km long, runs through southwest China and Myanmar to the Andaman Sea and is a treasure trove of biodiversity as already discussed.

The Ganges provides water for over half a billion people, more than any other river in the world. It is a sacred river for Hindus who believe it is the literal body of the goddess Ganga. The faithful believe that the purity of the goddess cleanses their sins and assists the dead on the road to heaven, so bathing in the river helps their sins be forgiven and allows them to attain salvation. As it flows 2,525 km from the western Himalayas to the Bay of Bengal, the river carries nutrient-rich sediment and deposits fertile soil along its shores.

The Brahmaputra River flows through Tibet, northeast India and Bangladesh, where it merges with the Ganges and discharges into the Ganges Delta. At 3,969 km long, it is a vital route for transportation as well as an important source of irrigation. Another sacred Hindu river, it is unusual in having a male name – Brahmaputra means 'son of Brahma'.

The Ganges remains one of the world's most polluted rivers with over two-thirds of sewage discharged into it being untreated. Due to industrial pollution of the river, it is estimated that around 20 million people in Bangladesh are exposed to water contaminated by arsenic so that they face the increasing risk of witnessing the cancer levels currently being experienced in China and India, again due to high levels of pollution.

India's damming of the Ganges has resulted in a diminished water flow downstream, especially where the Brahmaputra and Ganges merge in Bangladesh. Consequently, there has been serious damage to agriculture as a result of increased soil salinity. The social impact of this disastrous water management has been a huge migration of Bangladeshis into northeast India and thus a dangerous increase in the number of ethnic conflicts. The global impact of water events on migration is greatly underappreciated and will be discussed later.

The Mekong River, covering 4,350 km, is the seventh longest in Asia and the twelfth largest river in the world. The Mekong Delta, stable for thousands of years, has been massively disrupted in recent decades with its dams, waterfalls and rapids making the navigation of its entire length virtually impossible. In 1995, in response to common challenges they faced, the governments of Laos, Thailand, Cambodia and Vietnam formed the Mekong River Commission to co-ordinate the river's resources and maintain fish stocks and the habitat for other wildlife. Despite these noble intentions, the Delta's problems have not disappeared and it has been both shrinking and sinking in recent years due to the over-extraction of groundwater and the massive construction of hydroelectric dams, combined with the increased use of spring water upstream. It has produced the salinisation of groundwater, particularly coastal aquifers, along-side the destruction of rice crops, the depletion of wild fish stocks and the salinisation of soil. Unless drastic action is taken it is entirely possible that the Delta, as we know it today, may not exist in a century's time. It is possibly the single most horrific warning we have yet had about how lack of co-operation between neighbours and lack of governance upstream can have a fatal impact on communities and the environment downstream.

WHAT TO DO?

So, what is to be done? It is clear that co-operation and co-ordinated management is required if the region's stressed water supplies are not to result in human and environmental catastrophe or military conflict. Yet, distrust is rampant, particularly of China. The country's willingness to engage in rules-based co-operation is key, as is a change in mindset which stops its upstream position, through its control of Tibet, being used as a strategic weapon. Both its unparalleled dam building (understandable given its need for hydroelectric power and to supply water to the world's second-biggest economy) and its manipulation of the hydrological data it supplies to its downstream neighbours needs to change. If there is no change in attitude then we can expect major displacement and migration of populations, as well as the environmental impacts of pollution, deforestation, soil erosion and increasingly unpredictable flood patterns, each with their own unique and potentially irreversible costs.

On a wider political level, the lack of international legal frameworks to regulate transboundary water issues is a critical weakness. The Convention on the Law of the Non-navigational Uses of International Watercourses was passed by the UN General Assembly in 1997. As of mid-2023, twenty-four countries are registered as having full access to the treaty (the latest being Gambia in 2023), eleven have ratified the treaty and a further fifteen have merely signed it. Vietnam is the only country in the region to have acceded. There are a few brighter spots, however. India has recently reached an agreement with Nepal on barrages, dams and flood control. India and China have reached agreement on sharing some data relating to seasonal flood patterns for the Brahmaputra, and India also provides Pakistan flood forecast information.

While it is easy to focus on smaller downstream countries, such as those on the Mekong Delta facing all the dangers of China's insatiable damming projects, the United Nations Aquastat service suggests that of the surface water flowing out of the Tibetan Plateau from China's Xinjiang and Inner Mongolian provinces, over 48 per cent runs directly to India, which indicates the level of strategic advantage at which India would be placed should China's attitudes turn sour.

In 2017, in violation of two bilateral agreements, China refused to supply hydrological data to India with the consequence that early warning systems for extensive flooding failed and thus caused large numbers of, entirely preventable, fatalities. The excuse that China gave? That India had boycotted China's inaugural Belt and Road Summit, a clear example of how readily China might strategise its upstream water powers. Again in 2017, the main artery of the Brahmaputra River system, the Siang, became dirty and discoloured as it entered India from Tibet, raising the worry that China would treat its downstream neighbours with the same lack of concern it has for its own water quality and that pollution could become one of China's biggest exports to the region. Accusations were even made that this could have been a conscious act of environmental sabotage by the Chinese. Given India's considerable military capabilities, this is clearly a source of concern well beyond the region itself. It is perhaps the most likely point of conflict between two of the world's nuclear powers.

On the other side of India, disagreement with Pakistan over the Indus has resurfaced in recent months. India has demanded that

Pakistan modify the Indus Waters Treaty, which the two nations, plus the World Bank, signed in 1960. While the treaty, allocating the waters of several rivers in the Indus basin, has been widely seen as a practical and diplomatic success, the two sides have been making increasingly bellicose arguments about the use of this water since the mid-2000s. As so often occurs, the problem lies in the rights of the upstream and downstream partners. India wants to build hydropower projects on the Indus, the Jhelum and the Chenab Rivers. The treaty imposes limits on what India is allowed to do, and Pakistan is determined that India's dam projects fit with Pakistan's interpretation of the treaty, especially its own rights to make free use of these waters without upstream obstruction. It is a bad omen in one of the few areas where agreements on water use had, more or less, been made to work.

Then, of course, there are the effects of climate change to consider.

The World Bank has published research showing that the water availability in the region may impact on monsoon patterns as warming temperatures appear to be affecting the region differently from other parts of the globe. The biggest risk will be life-threatening floods in a region where, in recent years, Bangladesh, India and Pakistan have seen catastrophic flooding. Somewhat ironically, in the summer of 2023, China was hit by three Pacific typhoons in under three weeks, with the capital, Beijing, seeing its highest rainfall for 140 years. It shows that whatever sovereign powers national governments believe that they have, control of Mother Nature is not one of them. It is worth everyone remembering the effect that shifting monsoon patterns had on both the Mayan and Chinese civilisations in the past and imagining what similar shifts might do today.

A footnote. As if to sum up China's entire attitude on the issue of water in its region, in 2014 it established a $54 million development in Tibet to promote the bottled water industry and has now started to bottle Tibetan water. With all the pressures, conflicts, environmental hazards and potential impacts on the regional and wider climate, China has decided that one of the most important functions of Tibet's water is to bottle it and sell it!

CHAPTER 9

THE NILE

'I am Isis', Cleopatra famously declared. Ever the canny politician, in claiming to be the reincarnation of the goddess who presided over motherhood, the afterlife, healing and magic, she legitimised her position in Egypt, linking her with the eternal Nile and the ancient religions. The strategy also served her well in foreign policy, as she had no doubt intended. In Rome, she openly associated herself with the cult of Isis, helping to influence the construction of the first major Isis cult site at the Iseum Campense, and easing her path in what were complex Roman politics.

There are several myths associated with Isis, the most popular being how she grieved and sought justice for her husband Osiris, murdered by his jealous brother Set. Tricked into trying out a sarcophagus for size, Osiris was cast into the Nile. His wife eventually located his body with the help of children playing by the river, but when she returned with the body, it was dismembered by Set and scattered across the country. Isis eventually found all of his scattered body parts and built a shrine for each. All parts, that is, except his penis, which was eaten by a crocodile, an animal then associated with the god of fertility, Sobek, and so, the Nile was again a sign of the fertility of Egypt itself. Cleopatra clearly knew what she was doing in perpetuating the image that she chose.

Egypt developed early settlements on the banks of the Nile around 6,000 BC and evolved into what we would today regard as the first discernible nation state around 3,000 BC. It is hard to overstate the importance of the River Nile to the Egyptians. It was thought to be the source of all Egyptian life, with even the Milky Way thought to mirror its path.

At 4,132 miles (6,650 km), the Nile is the world's longest river and flows north from its sources to reach the Mediterranean Sea. Arising as the White Nile from Equatorial Africa, the Blue Nile from the Abyssinian highlands and the River Atbara which flows from northwest Ethiopia to the Nile and east Sudan, the river passes through Burundi, Kenya, Tanzania, the Democratic Republic of Congo, Rwanda, Sudan, South Sudan, Ethiopia, Uganda and Egypt itself. Today, these Nile River states have a population of around 400 million, with the UN estimating that this will double to over 800 million by the year 2050. While it is thought that around 10 per cent of this population currently face water scarcity, some studies suggest that this level could rise to as high as 35 per cent by 2040, although the fluctuating effects of climate change make this hard to predict with any certainty. It is thought, for example, that the current river flow of 84 million m^3 per year could vary by up to 50 per cent in the next fifteen to twenty years with unpredictable drought and flood conditions becoming more common, including the current dry conditions in the lower basin (Egypt and Sudan) as well as upper basin states which are much wetter today.

HISTORY AND GOVERNANCE

The legal management of the Nile today, hotly disputed by virtually everyone but Egypt and Sudan, is deeply linked to Egypt's history. Britain occupied Egypt in 1882 and maintained an extended and influential presence until the second half of the twentieth century, depending on Egyptian cotton for much of its textile industries. It was also key for ensuring safe and permanent access to the Suez Canal which maintained the crucial links with British India. It was understandable, then, that Britain was sympathetic to the demand for water rights claimed by the Egyptians, with British hydrologists helping to regulate the flow of the river by building reservoirs and dams.

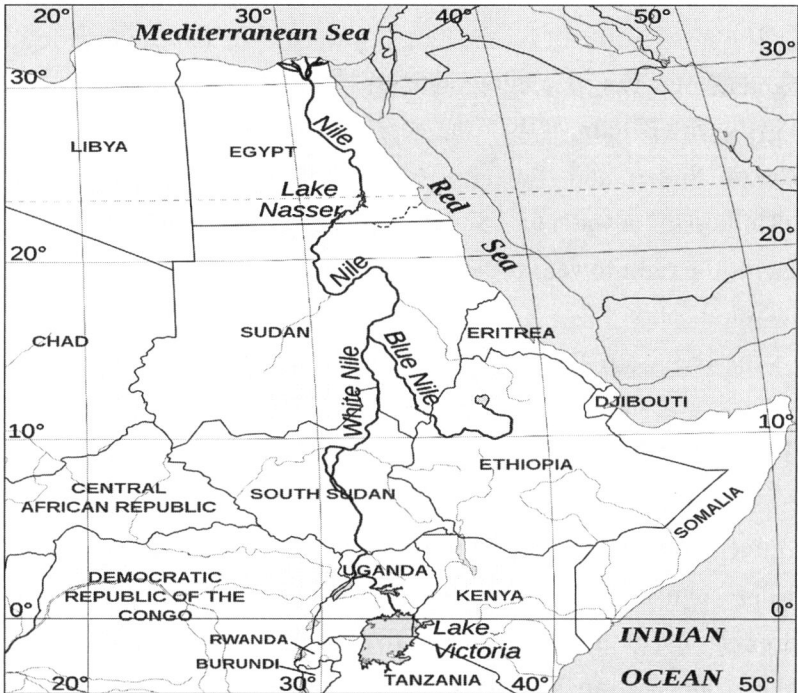

Egypt depends on the Blue Nile for 90 per cent of its water, so access to this was an essential strategic aim. If anything, that strategic imperative is even stronger today. In 1960, the water availability per capita in the country was 2,190 m³ per year. By 2019, it had dropped to 570 m³ and it is estimated to fall to a lowly 500 m³ by 2025. It is thought that Ethiopia's plan to build a huge dam upstream, the Great Ethiopian Renaissance Dam, would reduce by 25 per cent the flow of Nile water to Egypt during the dam's filling period (assuming that it would take around seven years) and that there would be a 30 per cent decrease in the energy production of the Aswan Dam over the same period. (Remember that less than 1,700 m^3 per capita per year is regarded as water stress, less than 1,000 m^3 is water scarcity and less than 500 m^3 is absolute water scarcity, according to the Falkenmark water stress indicator).

The distribution of Nile waters is governed by two existing legal agreements. The 1929 agreement was signed between Britain and Egypt, with Britain, at the time, representing its colonies of Uganda, Kenya, Sudan and Tanzania (then Tanganyika). It gave Egypt 48 billion m^3 of water and Sudan 2 billion. More importantly, it gave Egypt the right to veto projects higher up the Nile that might affect its water share.

The later 1959 Egypt–Sudan agreement supplemented the 1929 version and gave Egypt the right to 55.5 billion m^3 per year and Sudan 18.5 billion m^3 per year. In the years that have followed, Egypt has maintained that the two treaties, known as the Nile Waters Agreements, are the only legitimate international agreements that it will recognise. It continues to make the case that these agreements are required to prevent upstream nations from having an unacceptable strategic stranglehold on Egypt's water. Understandably, others take an opposing view. In 1961, following its independence from the United Kingdom, the leader of the newly created Tanzania, Julius Nyerere, claimed that the agreements left his country's water needs and development entirely at the mercy of Cairo's policy whims. It was hardly a meeting of minds.

In 1999, the NBI (Nile Basin Initiative) partnership was formed to increase co-operation, promote regional peace and better share the river's social and economic benefits. The Entebbe Declaration of 2018 went further, allowing the partner countries to build dams or related projects, superseding the terms of the previous colonial treaties. Unsurprisingly, Sudan and Egypt refused to sign it on the basis that Article 14 of the new agreement impinged on existing rights and water allocations, in other words, insisting that the colonial agreements were honoured. Despite an attempt by Tanzania to broker an agreement between Egypt, Sudan and the rest, no new plan that is acceptable to all has been reached.

Some argue, with justification, that the historic rights granted by the treaty have led Egypt to become over-dependent on the Nile for its water needs and therefore less able to be flexible in any negotiation. Yet, with its current population of 85 million expected to reach 140 million by 2050, Egypt's problems are real and acute. Once the grain house of the Roman Empire, half of the almost 20 million tonnes of grain that Egypt now consumes each year is imported, making it, in a massive historical turnaround, the biggest grain importer in the world.

When, in 2004, Tanzania announced its plans to build the Lake Victoria pipeline, which would have boosted supplies to around half a million of its citizens, Egypt threatened to bomb the construction site as it would have reduced the flow to the Aswan Dam, which produces over 10 per cent of Egypt's electricity generation.

All of this formed the backdrop to Ethiopia's decision to build the GERD (Grand Ethiopian Renaissance Dam), despite any agreement.

THE GRAND ETHIOPIAN RENAISSANCE DAM

The GERD, a US$5 billion project, now in its second filling phase, will be the largest hydropower project in Africa, creating a reservoir of 74 billion m³. Ethiopia argues that the project is key for its economic development and that it could lift millions out of poverty, as well as regulate the flow of the Blue Nile and reduce flood risks. With talks at an impasse (Egypt has failed to make headway via either the United Nations or the African Union), progress on building continued until, on 10 September 2023, the Ethiopian Prime Minister, Abiy Ahmed, posted on X (formerly Twitter) that 'it is with great pleasure that I announce the successful completion of the fourth and final filling of the Renaissance dam'.

At the heart of the dispute is the key doubt expressed by Egypt and Sudan about whether Ethiopia would willingly release sufficient water to downstream countries in the event of drought, a common phenomenon in the region. Under the chairmanship of the South African President, Cyril Ramaphosa, the three countries agreed that a reduction in the Nile flow to below 30 to 40 billion m³ a year would technically constitute such a drought. While the definitions were collectively agreed, the responses were not. Egypt and Sudan are demanding that under such drought conditions, Ethiopia must release more water, but Ethiopia wants to maintain flexibility in how and when it should respond. The crux of the dispute, as is so often the case, lies not just in the use of water itself, but how it intersects with nationalist politics.

Egypt believes the GERD is a strategic threat to its security as an independent state, while Ethiopia sees it as a legitimate development project within its national remit. Egypt sees the Nile as an integral part of its culture, history and identity as a nation, but Ethiopia sees the dam creating a new unity and challenging the country's international stereotypes as poor and backwards. Internally, the Ethiopian government promotes the dam as the centrepiece of a 'new Ethiopia-ness'.

Added to this, a narrative about how the project will 'Africanise' the Nile and threaten Arabian water interests proves once again that the water issue is seldom the single cause of potential conflict but is a potential trigger when there are pre-existing, competing concepts of national identity, culture and sovereignty.

How real, then, is the threat of open conflict? The rhetoric has been, like the grand old Duke of York, up to the top of the hill and down again. In 1979, Egypt's President, Anwar Sadat, declared that the only thing that would lead Egypt to war again would be the issue of water, particularly the Nile. President Morsi was even more bellicose, declaring that if his country's share of the Nile fell by one drop, then 'blood would be the alternative'.

Even now, the verbal seesaw continues. On 30 June 2020, the Egyptian Foreign Minister Sameh Shoukry warned the United Nations that Ethiopia's intention to fill the GERD would provoke 'crisis and conflicts' that would be a 'serious threat to international peace and security'. By 28 July, however, President Sisi was slapping him down, saying, 'Do not make threats and idle talk.'

So, what is the reality that underpins the rhetoric and what is the realistic probability that Egypt alone, or even aided by Sudan, could bring an end to the GERD project by military force? At first glance, the balance of power favours Egypt. In theory, it has a huge military advantage over Ethiopia. Its army is over 2.5 times larger. The EAF (Egyptian Armed Forces) have around 28,000 tanks, infantry fighting vehicles, Armoured Personnel Carriers (APCs) and Mine Resistant Ambush Protected vehicles (MRAPs), thirty-one times that belonging to Ethiopia. Egypt has nine times as many combat aircraft and possesses eighty-one attack helicopters compared to Ethiopia's mere eight. Furthermore, much of Ethiopia's equipment is of old Soviet-era capability, while Egypt has significant amounts of modern American weaponry, including F-16 fighters.

It is widely assumed that the most likely form of attack, were it to occur, would be from the air. In theory, Egypt could use its French Rafale fighter jets as well as its US F-16s but, in practice, however, Egypt simply lacks the operating radius to carry out such a mission successfully. Most of its bases are in the north, not the south of the country, and despite all their investment in air assets, Egypt does not have the necessary capability in tanker aircraft to enable air-to-air refuelling, something that would be required to make such an intervention a realistic option. Even if it were to be a viable operation, which parts of the dam would be attacked to make it inoperable, at least without leading to catastrophic failure and resulting in devastating effects downstream? In any case, although the Ethiopian Air Force is old and rickety by comparison, its air defences are not, with Israel supplying the state-of-the-art SPYDER system in 2019.

A land attack, often insinuated as a potential option, is also unrealistic. Again, Egypt lacks the southern bases from which to organise and launch such a mission. Even if such bases were available, it lacks the logistical ability to move troops. The GERD is over 800 miles from the Egyptian border, and it would require the permission of Sudan and/or Eritrea to move its forces across their sovereign territory. Such permission is unlikely to be forthcoming. It is, in effect, a non-starter. The same is also true of any special forces operation. Even if the difficulties of the targeting question could be answered, analysts seriously doubt that Egypt's special forces have the technical capability to carry out such an attack and far less to be safely removed afterwards.

It is also entirely unclear where external political support for any such operation would come from. Egypt's recent policy of so-called 'diversification' in foreign relations has been a mishmash of approaches, none with any notable success. While they have tried to get closer to Saudi Arabia and the UAE, both countries continue to have close links to the Ethiopian government, a position that seems unlikely to change. While Egypt has also tried to forge greater and deeper links to China and Russia, it seems to have overlooked the fact that China itself has made considerable investments in Ethiopia's wider electricity infrastructure. In fact, the rapprochement with China and Russia has largely had its greatest effect in irritating, and to an extent, alienating, Egypt's strongest ally, the United States.

The truth is that Egypt has never really possessed the ability to solve the GERD issue by force and does not possess such an ability today. Still, as of April 2023, naval forces from Egypt and Sudan have continued joint military drills from the naval base of Port Sudan on the Red Sea. Apart from a symbolic show of unity, it is difficult to see what impact it could have, in practice, against the GERD.

It seems then, that with no credible military threat against it, the GERD will ultimately be filled by Ethiopia. Yet, the implications of this will reverberate well beyond the immediate region and the countries directly affected. In the world of the upstream–downstream politics of water, if Ethiopia can complete its project and go on to fully generate electricity for itself without any agreement with its water-sharing partners, it risks sending a global signal that unilateralism is a viable and potentially successful strategy.

None of this will go unnoticed in Beijing or Moscow. If Ethiopia can set a precedent for unilateral action from an upstream controlling nation, what will it mean, given China's domination of Tibet and control over the vital rivers of South and Southeast Asia? It is little wonder that China has no appetite for helping Egypt find a solution mediated by the United Nations. It is the last sort of precedent that China wants to see. It is also worth noting the increased role that Russia and China are playing in the region itself. Both of those countries, for example, are big arms exporters, with 44 per cent of Egypt's arms now coming from Russia, who also supply 75 per cent of Ethiopia's military needs and 33 per cent of those in Sudan (where China also supplies 42 per cent).

With China establishing its first naval base in the region, the Horn of Africa seems set to be a potential flashpoint in global affairs. At the southern end of the Red Sea, the Bab-el-Mandeb Strait separates Yemen on the Arabian Peninsula from Djibouti and Eritrea in the Horn of Africa. With the Suez Canal at the northern end, it means that two of the world's most critical bottlenecks for trade will be permanently in play. The Red Sea is, therefore, only going to increase in geopolitical importance in the years and decades ahead. Add to this picture the construction of NEOM, the pet project of Saudi Arabia's effective ruler, Mohammed bin Salman Al Saud, or MBS, at the northern end of the Red Sea, close to the outlet of the Gulf of Aqaba, and the scene is set for a multinational tug-of-war.

The US Congressional Research Service (CRS) concluded in a 2020 report that 'failure to reach an accord could set a negative precedent for transboundary water cooperation at a time of growing global concern over climate change, demographic growth, and resource scarcity'. It will also provide needless obstacles to investment and necessary structural reforms.

Still, there must be hope that just as Israel, Jordan and the UAE were able to use their respective strengths to reach the agreement to swap water for energy, so the nations affected by the GERD can do the same. If, for example, the regional neighbours were to establish a properly integrated power grid and solar and wind energy alongside the hydroelectric potential of the Nile, then the benefits might be astronomical.

In a powerful contribution to the debate, John Mukum Mbaku of the Brookings Institute, believes that a way forward can be found if all the countries involved (Burundi, Democratic Republic of Congo [DRC], Egypt, Eritrea, Ethiopia, Kenya, Rwanda, South Sudan, Sudan, Tanzania and Uganda) start to focus on mutually beneficial issues such as trade, the prevention of terrorism and extremism and confronting the common threats to greater prosperity, such as climate change, widespread illiteracy and poor infrastructure. 'It is vital', he says, 'that all the upstream states must recognise and accept the almost total dependence that Egypt has on the Nile waters for its existence, while, at the same time, Egypt must abandon its constant references to its so-called "natural historical rights".'

Any substantial agreement, not to mention the practical mechanisms of enforcing it, such as dispute resolution procedures, still seems far away. Yet, the fact that the issues cannot be determined by military might means that diplomacy is ultimately the only viable route to a workable agreement. Only visionary political leadership, leaving aside partial national interests, can ensure that the people who depend on what is probably the world's greatest river for their very existence can have a fair, equitable and safe future.

CHAPTER 10

IRAN

Most global disputes over water have a predictable list of ingredients: the demand on natural resources, human mismanagement and concepts of sovereign rights. When it comes to Iran, and particularly its dispute with its neighbour Afghanistan, we need to add political toxicity, religious extremism and systematic brutality, all allied with chronic incompetence. Iran's clerical government, one of the most malevolent and oppressive in the world, is not your average 'thugocracy', governments where brute force and criminality oppress the population. Iran goes further. It is a 'thugocracy' (enforced by the Islamic Revolutionary Guard Corps [IRGC], the Revolutionary Guard answerable only to the Supreme Leader) with an extreme theocratic dictatorship on top. Its human rights abuses are legion and its use of detention, torture and execution are indispensable tools of one of the world's most sordid regimes. The lack of attention that these crimes attract from parts of the western media leads many of us to wonder whether Iran has been adept at placing, or purchasing, those sympathetic to their interests in positions of influence. When it comes to their neighbours, the Taliban in Afghanistan, they face stiff competition for the title of the most appalling, most anti-women and most medieval government on the planet. Since the west's self-induced humiliation in the withdrawal from Afghanistan, the position of its citizens, especially women, has steadily worsened, with an increasingly brutal denial of basic liberties. A style of government that would probably have been regarded as backwards in the thirteenth or fourteenth centuries is now increasingly, and viciously, being imposed upon the people of Afghanistan in the twenty-first century. Now, add to this septic political mix with its built-in disregard for basic humanity, the issue of water, with all its complexities.

In early 2023, Iranian officials confirmed that 270 cities and towns were in a state of acute water shortage with dam levels dropping dramatically. The authorities' response was typically draconian. There was no hint of responsibility for decades of mismanagement of the water infrastructure, but a threat to raise water prices 'to correct the consumption pattern'. In other words, the mullahs believe that the people are at fault for this problem, not them. By the middle of the year, record heat and humidity exacerbated the difficulty and the belief that the water shortages have been driven by decades of appalling negligence has led to discontent and open protest. In July, the head of the IRGC visited the chronic drought-stricken province of Khuzestan but warned against any demonstrations during his stay. This reflects the increasing nervousness of a regime which provoked outrage following the death of the 22-year-old woman, Mahsa Amini, who died following brutal treatment from Iran's so-called morality police. The desperate crackdown by the clerics and their agents, using detention and even execution to try and suppress public demonstrations has led to even greater hatred towards them among the Iranian people, especially young people and particularly women. As ever, the political tactic is to divert attention, and the regime has tried to blame the Taliban and the long-standing feud between Iran and Afghanistan over the

Helmand River for their own water issues. The truth, however, is that the regime has constantly sought to pacify its supporters in the more backwards and rural parts of Iran by ensuring the diversion of water to the agricultural sector. Currently, more than 93 per cent of the available water in Iran is used in agriculture, leaving only 7 per cent for industrial use or drinking water. Perhaps the most worrying element of a dreadful picture is the huge drop in the level of groundwater, where centuries of water laid down in aquifers is being depleted at alarming rates.

THE TRIALS OF KHUZESTAN

The second most economically important province in Iran, Khuzestan, sits in the southwest of the country, with a population of almost 5 million people. Despite having over one-third of Iran's surface water resources, water access for the people of the province has been a serious and chronic problem. Although it provides 15.5 per cent of the country's GDP, poverty there is endemic, unemployment is high and eleven cities in the province are in drought with 700 villages having no potable water.

The water stored in the province's aquifers is estimated to amount to around 3,000,000,000 m³ and the water basin is home to some of Iran's key rivers. The Karun, which has the highest flow of any Iranian river, flows 590 miles from the Zagros Mountains, through Ahvaz, the capital of Khuzestan Province, before emptying into the Shatt al-Arab. The province is also home to the Shadegan and Hur Al Azim waterways and numerous dams that are designed to transfer water to other regions. This is the key to many of the problems suffered by the population of Khuzestan.

In October 2018, parts of the province, particularly the districts of Abadan and Khorramshahr, faced severe water shortages that resulted in most of their agricultural lands being destroyed. In the years that followed, hot summers led to high levels of evaporation from the poorly managed dams, especially the Karkheh Dam. The situation was greatly worsened by the continued diversion of the Karun River in order to supply agricultural land in areas where the rural population was a crucial element in the political support afforded to the Iranian government. Needless to say, Khuzestan, with a large ethnic Arab population, is a lesser consideration for a deeply fundamentalist Shia regime that is steeped in its Persian historical and cultural identity. The dysfunctional nature of the disjointed government continues to add to the problems, with unco-ordinated activity between the Ministry of Agriculture, the Ministry of Energy and the Ministry of Petroleum all jostling for political and economic dominance.

In June 2020, protests in different parts of the province erupted almost simultaneously, with people comparing their own plight, in a technically water-rich part of the country, with the much more plentiful supplies enjoyed by the residents of the capital, Tehran. The demonstrations were put down with characteristic violence, an act of particular significance and symbolism in a part of the country where many fought and died during the Iran–Iraq War as patriots. To have their own troops firing on them as they tried to obtain the most basic of commodities stoked their anger and guaranteed rising resentment.

A year later, in July 2021, protests broke out across the province again, driven by the mass death of livestock and the consequently almost impossible task of eking out a living from the parched land. This time, the protests spread with unprecedented pace and, more worryingly for the Iranian authorities, almost all the cities of Khuzestan were affected and the protests were mirrored in other cities, such as Isfahan, Kermanshah and even parts of Tehran. Now, the repression by the security forces was even more severe than before, with water cannons and armoured vehicles being moved from the capital to the provincial cities. As the authorities shut down the internet in a bid to silence dissent, many in the international community heard demonstrators calling, for the first time, for 'death to Khamenei' and 'death to the Islamic Republic', calls that have become more familiar in recent months as a growing number of disputes have laid bare the division between the people and the religious and social repression of the mullahs. On 23 July, in a classic piece of stage-managed deception, the Supreme Leader appeared to sympathise with the demonstrators, saying that 'no one can complain about these people. If the necessary measures [on waste removal and water shortages] had been adopted … this situation would definitely not have arisen…. The problem of water is not a minor issue … particularly considering the exemplary loyalty of the people of Khuzestan.' Meanwhile, in an act of breathtaking hypocrisy, the forces under his own direct command attacked these 'loyal' citizens. In one, now famous, recording, a woman called out to the police, 'These demonstrators are peaceful. Why are you shooting? They haven't taken away your land, have not taken away your water. We are protesting peacefully.' The recording ends with gunfire ringing out.

In February 2022, residents in parts of the city of Ahvaz found themselves without piped water for ten consecutive days and by the summer the continued diversion of the Karun River saw the Kharkeh River suffer from an acutely reduced flow, while the Dez River totally dried up downstream. Widespread migration from the countryside to the cities gathered pace and was exacerbated by rising food prices imposed by central government and the dust storms created by the environmental disaster. In some cities, such as Ahvaz, the security forces now maintained a 24-hour presence to control the 'loyal' citizens and an increasing number, which will probably never be known exactly, were detained or disappeared. Meanwhile, the authorities continually promised to resolve the water situation but, in July 2023, when the residents of Sayyahi, south of Ahvaz, complained that for twelve months their water had been cut off, the pressure was low and it stank of sewage, the CEO of the Khuzestan Water and Sewage Company blamed the increased population for the problem. He insisted, however, that the problem of the water shortage would be fixed 'in five days'. It remains unresolved today.

Khuzestan has abundant reserves of oil and gas. It has thirty-three petrochemical facilities, four huge oil refineries and manufacturing capacity for steel and iron. It plays a key role in transporting and exporting Iran's energy products. It also has a large natural water basin. Yet, despite their major role in the Iranian economy, the province's population sees that while they endure perpetual water shortage, water pipelines have reached the area for the camps that house the oil and gas workers.

In Gheyzaniyeh, a town of 11,700 mostly agricultural workers, the residents find themselves surrounded by gas and oil pipelines with no water pipelines of their own. Worse, they complain that the industrial pollution of the Karun River is damaging the quality of what water they are able to get, with one commenting that 'they process the colour out and give it to people as drinking water'. According to the Deputy Provincial Director of Environmental Protection in Khuzestan, '70 per cent of the wetland in the province has been destroyed by the contractors of oil companies. Reservoirs three, four, and five of the wetlands were kept dry for the oil facilities to carry out their activities, and water was not allowed to enter these areas'. In 2016, the Governor of Khuzestan promised that the water problem would be solved 'in three months'. In 2021, he claimed that 'the supply of drinking water for the villagers in this district will be secured for the next twenty years and they will no longer have any drinking water problems'. Three and a half years later, the water pipes of the residents of Gheyzaniyeh are as empty as the promises made to them by the Iranian authorities.

Even in a state so rigorously and violently repressed as Iran, it is doubtful whether the people's grievances, political, economic, religious and environmental, can be suppressed indefinitely. What seems certain is that the ruling theocrats see their authority being increasingly threatened and their privileged positions in jeopardy. History suggests that their response will be correspondingly brutal, with their primary concern being for their own survival and their responsibility for the needs of the Iranian people non-existent. Yet, while the domestic picture deteriorates, pressures are also mounting outside Iran's borders.

IRAN'S CLASH WITH AFGHANISTAN

The Trump administration agreed to withdraw American forces from Afghanistan by 1 May 2021, if the Taliban negotiated a peace agreement with the Afghan government and promised to prevent Al Qaeda, Islamic State and a number of similar groups from gaining a foothold in the country.

When the chaotic withdrawal of forces eventually took place in August 2021, a rapid Taliban offensive brought about the fall of Kabul and the beginning of the tyranny and human rights abuses that now afflict Afghan civilians. Many of us saw the entire episode as a betrayal of all the sacrifices made by the international coalition to not only bring freedom and prosperity to ordinary Afghans but also reduce the risk to the international community from Islamic extremists.

Since the Taliban's return, hundreds of thousands of people, mainly women, have lost their jobs. In their first press conference after taking power, the Taliban announced that 'we are going to allow women to study and work within our framework. Women are going to be very active in our society.' Predictably, it now seems that 'the framework' is fundamentally anti-women. A month after the announcement, girls were barred from secondary schools. Three months later, travel restrictions were imposed and when women took to the streets in protest, many were held for weeks and beaten in custody. In May 2022, new dress codes were imposed on women, followed by bans on attending university, using public spaces and non-governmental organization (NGO) work. In July 2023, beauty salons were closed, further reducing the possibilities of employment for women and, in August 2023, they were banned from one of the country's most popular recreation parks. It seems hugely ironic that at a time when, in the west, an inappropriate kiss at the women's football World Cup final can occupy acres of media copy, the appalling treatment of women in Afghanistan, a direct consequence of western policy, receives relatively little attention.

As far as the Iranians are concerned, the net result of the Taliban's return is that they are now in control of the Kamal Khan Dam whose completion reignited a dispute between the two countries over the use of the Helmand River's water resources. When the dam was inaugurated on 24 March 2021, even before the Taliban takeover, President Ghani set the tone by declaring that 'Afghanistan would no longer give free water to anyone, so Iran should provide fuel to Afghans in exchange for water'. It was, perhaps unsurprisingly given the nature of the two countries, the most brutal example yet of how an upstream nation might explicitly use blackmail over water as a political tool.

The Helmand River Water Treaty of 1973 gave Iran 22 m³ of water per second, with an additional ability to buy 4 m³ per second in 'normal' water years as an act of 'goodwill and brotherly relations'. Other provisions of the treaty effectively gave Afghanistan unilateral rights over the rest of the water supply for hydroelectric generation, reservoir, agricultural or industrial purposes, with their sole responsibility being not to pollute the river water.

A parallel tension to that which endures over the Helmand River exists over the Harirud River, which flows from the mountains of central Afghanistan, through Iran's northeastern territories and on to Turkmenistan. The Salma Dam on the river, renamed the Afghan–India Friendship Dam, thanks to the funding and construction by India, was opened on 4 June 2016 by President Ghani and the Indian Prime Minister Narendra Modi. Iran has twin fears: that its farmers in the northeast of the country might be vulnerable to Afghan water restrictions, while Afghanistan's increased ability to generate hydroelectric power might make it less dependent on Iranian energy and remove, or at least diminish, one of Tehran's strategic levers.

There may, however, be more than just bullying at play in Afghanistan's intransigent behaviour. Many believe that a thirty-year gap in knowledge and practical experience has left Afghanistan poorly trained in international law and lacking negotiating skills, leaving them fearful about being outmanoeuvred by their more sophisticated Iranian counterparts. There is also a suspicion about Iran's typical duplicity since the Taliban will be well aware of Iranian attempts to encourage the disruption of Afghan water projects while the previous government were in office.

Yet, both sides must carry some blame for the current parlous state of affairs. The Hamoun Lake was once the third-largest lake in Iran, fed by Helmand River water. It was a crucial water source for those living in Baluchistan and the southeastern Sistan provinces.

Now, a combination of factors, including canal creation, inappropriate dam building and the diversion of the Helmand River to artificial reservoirs have left it largely dried up. The Hamoun wetland also had its entire vegetation cover destroyed when inappropriate, invasive fish species were introduced for commercial reasons in the 1980s.

All of these problems have been made worse by climate change, with temperatures in Afghanistan having risen by 1.8°C since 1950, and where the incidence of both drought and excessive rains have increased, bringing even more misery to the Afghan people.

Despite the rise in unemployment, crippling international sanctions, a hunger crisis and being cut off from the global financial system, the Taliban continue to increase tensions with their neighbours over water issues. As well as their disputes with Iran, the Taliban is also building a canal to divert water from the Amu Darya basin that would flow into Uzbekistan and other central Asian states.

It seems that the two toxic regimes of the Iranian mullahs and the Taliban will continue to impose misery, incompetent management and religious extremism on their domestic populations and ramp up external tension and potential conflict for the foreseeable future. Their shared combination of religious extremism and political repression can only bring more hardship to those unfortunate enough to live under their rule.

CHAPTER 11

AFRICA... AND ELSEWHERE

To say that Africa is a continent of extremes would be an obvious understatement. Containing the world's largest hot desert, the Sahara, with an area of 3.6 million mi² (9.2 million square kilometres) and the world's oldest desert, the Namib, which has suffered from arid conditions for between 55 and 80 million years, it has its share of the world's harshest dry environments.

These are in sharp contrast to the lush tropical and subtropical wet regions which include some of the world's mightiest rivers, greatest waterfalls and most magnificent lakes. The continent's total renewable water resource is thought to be around 8,200 km³ per year, which is around 14 per cent of the world's total. The water supply is, however, grossly mismatched to the population, with around 70 per cent of renewable water found in central and western regions where only 34 per cent of Africa's population live. According to a United Nations press release, 500 million people living in nineteen African nations are deemed 'water-insecure'. In 2022, access to drinking water ranged from 99 per cent in Egypt to only 37 per cent in the Central African Republic. Therefore, around 29 per cent of the total population of Africa lacks basic drinking water, which amounts to some 353 million people. More than 2 billion people live in high water stress, a situation made worse by poor quality (often corrupt) governments, low investment, rapid population growth (particularly in overpopulated centres) and changing weather patterns due to climate change.

THE GREAT LAKES AND RIVERS

The African Great Lakes are part of the Rift Valley lakes in and around the east African Rift. Collectively, they contain 31,000 km³ (7,400 cu mi) of water, more than the North American Great Lakes. Their water content accounts for around 25 per cent of all the unfrozen surface freshwater on the planet. Lake Victoria is regarded as the second largest freshwater lake in the world, with an area of around 68,800 km², although it is relatively shallow. Lake Tanganyika is the world's second-largest freshwater lake by volume and depth and Lake Malawi is the eighth-largest freshwater lake in the world by area and is the third deepest lake on earth.

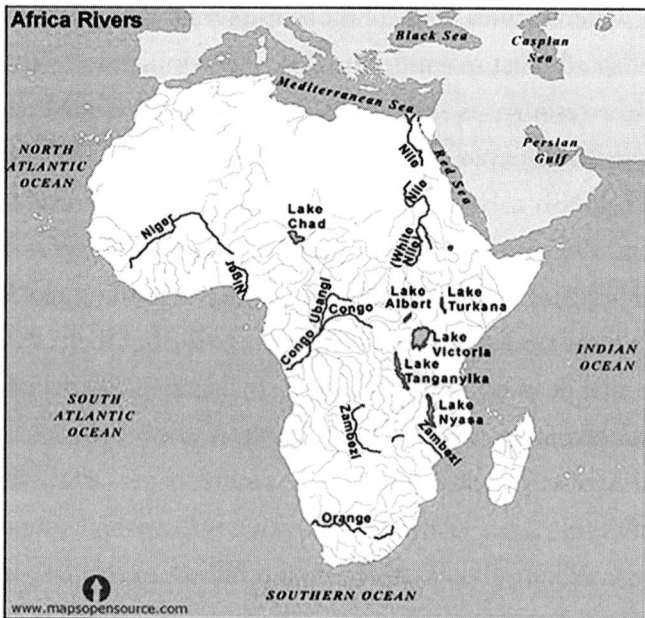

The Nile we have already considered in detail, but Africa has many other great rivers.

The Congo comes second to the Nile in terms of length, flowing all the way from equatorial east Africa to the Atlantic Ocean in the west, crossing the equator twice, the only major river that does so. While today it is a major source of hydropower, it was once christened 'the heart of darkness' due to its associations with the ivory and slave trades and intertribal warfare. The title is immortalised in the book by Joseph Conrad in which the sailor Charles Marlow tells the story of his adventures as a steamer captain for a Belgian company in the African interior. Probably the best-known film version of *Heart of Darkness* is, in fact, *Apocalypse Now*, though the story was transplanted from the nineteenth-century Congo to the war in Vietnam.

The Niger is the third-longest river, originating in Guinea and crossing west Africa until it empties into the Gulf of Guinea in Nigeria. Consequently, its main source is a mere 240 km from the Atlantic. It plays a key role in providing water for agriculture in the Sahelian region and western Sahara owing to its numerous dams and irrigation channels.

The Zambezi originates in Zambia and crosses Angola, Zambia, Namibia, Zimbabwe and Botswana, emptying into a wide delta in the Indian Ocean in Mozambique. It is home to the Victoria Falls, located on the border between Zambia and Zimbabwe. Identified and named by the Scottish missionary David Livingstone in 1855, it is one of the world's largest waterfalls with a width of 1,708 m (5,604 ft).

The Senegal River, which flows for around 1,800 km from west to north, is in fact the convergence of two rivers, the Semefe and the Bafing, both of which originate in Guinea. For much of its length, it marks the border between Senegal and Mauritania and has been a source of friction between the two countries. The river is depicted on the sea chart by Mecia de Viladestes who set new standards of cartography in 1413, including for the depiction of the Sahara desert. The river is thought to be that depicted as the 'river of gold' as, he commented, 'the greater part of those that live here occupy themselves collecting gold on the shores of the river which, at its mouth, is a league wide, and deep enough for the largest ship of the world', making it an attractive signpost for traders of the day.

South Africa has two of the main rivers – the Orange River and the Limpopo River. The latter, 1,600 km long, is the most important, providing life-giving waters to Zimbabwe and Botswana (where the river provides the border with South Africa) before it empties into the Indian Ocean.

Despite having enormous variability in their flows, between seasons and years, as a result of variations in patterns of precipitation, many of the rivers are navigable and, thus, useful sources of trade and transport. They are also, of course, vital for human water consumption, sanitation and electricity generation. The rivers can also be a potential source of disputes between, and sometimes within, states when the competition for water becomes more extreme.

AFRICAN WATER-DRIVEN CLASHES

The Senegal River basin has seen open conflict between groups divided on ethnic grounds and also between the states of Mauritius and Senegal, both former French colonies which have developed in different directions politically. For centuries, the basin has been inhabited by indigenous tribes, such as the Fulani and the Soninke and later by Arabs and the Berber people. Between 1989 and 1991, conflict around grazing rights erupted between Fulani herdsmen and

Mauritian Soninke farmers. Again, as with many disputes, notions of identity and sovereignty added to the tension, with Mauritania reinforcing its Arab identity following independence and Senegal maintaining its Francophone orientation. As riots broke out and the military intervened, both countries began expelling the other's nationals at the end of April 2019. Hundreds died, around 160,000 Mauritanians in Senegal were repatriated and an estimated 250,000 people fled their homes as a result of cross-border attacks. Both sides were to blame for the mass disruption of people, which saw a huge decline in agricultural production and increasing deforestation and, as if to underline the futility of the dispute, the Mauritanian construction and fishing industry, traditionally manned by Senegalese workers were heavily hit. The legacy of the dispute, in terms of antagonistic relations between the two countries, continues today, providing an unnecessary hurdle to economic and social progress.

Senegal and Mauritania are in no way alone. There are long-standing tensions between Ethiopia and its neighbour Somalia which go back to the determination of the border in 1948, which left many Somali nomads in the Ethiopian Ogaden desert. Disputes over the ownership of the desert water led to the Ogaden War in 1997–98 and today, Somali groups threaten any Ethiopian or other settlers who do not have their authorisation. In 2007, nine Chinese oil engineers were kidnapped and murdered, a clear sign that the dispute does not recognise 'neutral foreigners'.

As Lake Chad began to lose its surface water, dropping from 22,772 km^2 in 1966 to 4,398 km^2 in 1975 and only 1,756 km^2 by 1994, Nigerian fishermen moved to Cameroon, sparking military clashes between the two nations who, again, had a dispute of borders, post-independence. Despite the International Court of Justice (ICJ) ruling in 2002 that the territory in the Lake Chad region should be under the sovereignty of Cameroon, tensions persist to this day.

There have also been violent clashes between Kenyan and Ugandan herders, such as those which occurred in 2012, when Kenyan herders crossed the border looking for water and pasture and thousands of Ugandan troops were sent to control the unrest between rival populations. In the same year, clashes also occurred on the border between Mali and Burkina Faso when an agreement on sharing water and pastures was revoked and the nomadic Fulani and Dogon ethnic groups attacked one another. The dispute continues today, with the most recent fatalities occurring in March 2019 when 160 people were killed in a Fulani village. It is yet another of the continent's ongoing disputes relating to water, which achieve very little coverage in the western media, but which are a stark warning about the risk to security and stability in the future.

Somalia is a state worth looking at in a little more detail as it represents several different facets of the water problem within a single country. Here, security and developmental challenges go hand in hand and the federal government, established in 2012, has struggled to create a functioning state against the backdrop of drought, famine and armed opposition from the extremist Islamic group, Al-Shabaab.

According to recent estimates, around 4.2 million people in Somalia face severe water shortages. Three decades of civil war have had a devastating impact on water systems and more than 3 million people have been internally displaced. Lack of effective governance and regulation, compounded by increasing strains on groundwater resources and rising water prices, have worsened the Somalian people's already miserable state. The long distances that many people must travel to access water has resulted in the overuse of ground wells, contaminating the groundwater that contributes 80 per cent of the country's resource. The sanitation nightmare caused by open defecation, pit latrines and humans and livestock sharing the same water points is all too prevalent in a water system that is increasingly, and dangerously, polluted.

As if this situation were not dreadful enough, Somalia has also become vulnerable to the increasing severity of the droughts and severe rain events that seem to be affecting the Horn of Africa as the effects of climate change start to bite.

Recurrent and severe droughts have seen a catastrophic destocking of herds, often the only option available when this is the only asset that individuals have to sell. The problem is that when this occurs on a widespread scale, prices can drop drastically as local markets become oversupplied with livestock that nobody wants to buy. Added to this, a number of Somalia's main export markets, largely in the Gulf, have restricted or banned livestock imports on public health grounds. The economic knock-on effect has been the falling value of the Somalian shilling which has raised the import prices for basic commodities like flour, rice and fuel.

It is little wonder that the net effect of these combined horrors has been an increasing dependence on relief aid in camps where political extremists have been quick to recruit those who can earn money from illegal activities such as livestock reading, or even piracy.

There is a clear lesson here, which has been repeated across the globe in various circumstances, that when people are unable to achieve prosperity through legal means, including trading their way out of poverty in a free-trade system, they are likely to turn to other means of saving themselves and their families. In an interconnected world, if we are unable to help solve some of the economic problems that afflict poorer nations, we should not be surprised if we see more mass migration or more political radicalisation as a result.

All the problems faced by the Somalis, and reflected in many other water-stressed states, will be worsened if drought and flood become more frequent because of climate change. The consequences will be felt well away from the source of the problem as the effects ricochet around the globe in one form or another.

EVEN IN CALIFORNIA...

Lest we think that water problems are only the preserve of developing countries, let's consider a salutary tale of how even the most prosperous part of the world's most affluent and powerful nation can be affected. The California Water Wars, as they have come to be known, were a series of conflicts, physical and legal, between farmers and ranchers in the Owens Valley of eastern California and the city of Los Angeles, all centred around water rights.

Originally, the Owens Valley was inhabited by the Paiute Native Americans. In the second half of the nineteenth century, as ranchers started to raise cattle on the rich pastures of the valley, most of the Paiutes disappeared, courtesy of the US Army during the Owens Valley Indian War of 1862–63. By the end of the century, most of the land had been claimed by those from the east seeking a new life and, while the south of the country was mostly used for rearing livestock, the rich system of irrigation in the north of the valley was producing poultry, dairy and fruit as the twentieth century dawned.

Around the same time, the city of Los Angeles began to develop. Whether or not Los Angeles is, in fact, part of a desert environment has long been a bone of contention. Many would argue that a true desert will get less than 10 inches (25 cm) of rain per year and the rate of evaporation will exceed the rate of precipitation. On average, Los Angeles gets almost 15 inches of rain per year, though in recent years it has been as little as 5 inches. So, it can be reasonably argued that if it is not a genuinely desert climate, then it is certainly on the cusp. As the city grew, the Los Angeles Water Company realised that the Owens Valley, with a large runoff from the Sierra Nevada, could provide a vital water resource for the mushrooming city if only a workable system of infrastructure could be constructed. The idea that was settled upon was the building of an enormous aqueduct. Before this could happen, however, Los Angeles needed to procure rights-of-way across federal land, a process that became highly controversial, being settled in the end by President Roosevelt, who sided with the city against the farmers.

The Los Angeles Aqueduct was inaugurated in November 1913, and as the city's demands for water grew, more and more was taken from the Owens Valley until the once rich and fertile ground became impossibly dry for farming, with the Owens Lake finally drying up by 1924. Understandably, farmers and ranchers rebelled, dynamiting part of the system and allowing water to return to the Owens River. Unfortunately for their cause, because of criminal embezzlement, the local Inyo County bank collapsed and, with it, the ability of local farmers to maintain their resistance against LA's rich and powerful leaders. By 1928, Los Angeles owned 90 per cent of the water in the valley, leaving its agricultural activities as part of its history.

This, however, was far from the end of the story for the Owens Valley. In 1970, the Los Angeles Department for Water and Power (LADWP) built a second aqueduct and so severely depleted the valley's groundwater that once widespread springs disappeared, the earth began to dry and ground vegetation withered and died.

Despite long-running court battles and attempts at a negotiated settlement, groundwater continues to be pumped out today at a rate higher than the aquifer can be replenished, setting the Owens Valley on a seemingly irreversible route to desertification, while the gardens of Los Angeles continue to be watered.

In this valuable tale, the real losers are not the farmers and ranchers of the Owens Valley themselves, terrible though their losses have been. The real loser is the environment, where the natural resources of ground and surface water were used with no regard for the long-term consequences. It is a story that is an increasingly recognisable trend around the world. Increasing urbanisation, particularly to support such a large human population, will come at an unavoidable cost to the environment. When the inevitable shortages occur, even when technology such as desalination can mitigate the damage, political tensions are likely to rise. When these tensions become intermingled with concepts of ownership, or even sovereignty, the possibility of conflict will, unavoidably, become all too possible.

CHAPTER 12

MARITIME CHOKEPOINTS

One final area in which water plays a crucial role in global economics, geopolitics and security relates to maritime chokepoints. These are strategic, narrow, sea passages that connect two large areas to one another, usually straits or canals experiencing high volumes of traffic because of their key locations. They can be affected by structural problems due to accidents or mechanical damage, or they may be deliberately targeted in geopolitical disputes.

With around 80 per cent of global merchandise being shipped by sea, commercial shipping relies on strategic trade routes to move goods efficiently and supply the constant global clamour for more consumer items. These trade routes can cut thousands of miles and days from travel times but do not come without risks. While they bring increased efficiency they also, almost inevitably, diminish the overall resilience of the system.

The sort of structural risks that are associated with narrow passages was amply demonstrated recently in the Suez Canal blockage. In March 2021, the canal was blocked for six days after the grounding of the 20,000 twenty-foot equivalent unit (TEU) container ship *Ever Given*, following a sandstorm where winds exceeding 74 km/h (46 mph) resulted in an inability to steer the ship, whose hull deviated. The obstruction by the vessel, the length of four football pitches, occurred south of the section of the canal that has two channels, so there was no way for other ships to bypass *Ever Given*. Consequently, by 28 March, at least 369 ships were queuing to pass through the canal, preventing the movement of an estimated US$9.6 billion worth of trade. The ship was finally freed by Egyptian, Dutch and Italian tugs and the canal was found to be undamaged, but the blockage temporarily sent oil prices climbing on the international markets, a sign of the immediate economic impact that closure of a chokepoint could have.

Chokepoints are also subject to geopolitical risks, either through direct targeting by insurgents or because they pass through dangerous or contested territory. Despite these risks, the financial rewards mean that they remain key waterways in the facilitation of international trade. Some of the most contentious and vulnerable of the primary chokepoints are:

STRAIT OF HORMUZ

Effectively controlled by Iran and connecting the Gulf to the Gulf of Oman and the Arabian Sea, this is probably the most important chokepoint in the world. Around 21 million barrels of oil pass through here every day, about 30 per cent of all the oil traded on the world's oceans, literally a vital lubricant for the world economy. It has long been an area of conflict with tankers and commercial ships being attacked during the Iran–Iraq War in the 1980s and, more recently, increased harassment of international shipping by Iran. With the election of the new hardline President, Ebrahim Raisi, this is a situation that is unlikely to improve. On 4 January 2021, Iran's armed seizure of a South Korean tanker, the *Hankuk Chemi*, added to the increasingly long list of incidents in which civilian ships have been deliberately targeted. Iranian forces boarded the vessel in the

Strait less than a week after another tanker, *Pola*, was attacked with a limpet mine. These are the latest in a long line of high-profile incidents, including the seizure of the British tanker *Stena Impero*, in 2019. According to the Centre for Strategic and International Studies in Washington (CSIS):

> Iran is reshaping its military forces to steadily increase the threat to Gulf shipping. This increase in Iranian capability is almost certainly not designed to take the form of a major war with the US and Southern Gulf states, which could result from any Iranian effort to truly close the Gulf. It does, however, give Iran the ability to carry out a wide range of much lower-level attacks which could sharply raise the risk to Gulf shipping.

Even if Iran were unable to stop the flow of oil through the Strait, consistent lower-level attacks could have a major impact on insurance costs and global oil prices. More worryingly, given the regime's intransigent nature, its attempt to destabilise its neighbours and its willingness to use international terror groups as proxies, there seems to be no cause for optimism that the situation will improve any time soon. As Iran tries to move inexorably towards becoming a nuclear state and with the increased co-operation it has received from Russia in its ballistic missile programme (as a reward for supplying Iranian drones in the war against Ukraine) the risks all look to be increasing.

THE SUEZ CANAL

The Suez Canal in Egypt is a long-standing geopolitical prize that links Europe with Asia. Its absence, or lack of availability, would add around seven days to maritime journeys if ships were forced to sail around Africa. In 2019, almost 19,000 vessels with 1 billion tonnes of cargo (including around 4 million barrels of crude oil per day, along with refined products) sailed through the canal. The passage has recently been gaining importance as a southbound route for US and Russian crude oil and petroleum products to destinations in Asia and the Middle East. Crude oil shipments, mainly to Asian markets such as Singapore, China and India, have more than doubled in the past three years with petroleum exports from Russia accounting for the largest share (24 per cent) of southbound petroleum traffic. We will see, in due course, what effect global sanctions in the wake of Putin's invasion of Ukraine have on these flows. In the past two years, increased production and exports of US crude oil and petroleum products – especially liquefied petroleum gas – have also increased southbound traffic through the canal. Despite the Egyptian government's major expansion project for the canal in 2015, risks remain as the episode of the *Ever Given* proved.

THE PANAMA CANAL

Of all the places I have visited, this remains one of the most impressive. The magnificent Panama Canal provides a shortcut between the Pacific and Atlantic oceans, saving a trip of around 8,000 nautical miles, equivalent to around twenty-one days. The canal underwent a $5.4 billion expansion in 2016 which tripled the size of ships that could pass through it. In 2019, over $2.6 billion in tolls were generated, while in 2020, 12,245 transits carried 255.7 million tonnes of cargo through the canal. Of these, 2,759 were liquefied natural gas (LNG) transits, 1,305 carried petroleum gas and 712 carried crude products. The increasing presence of Chinese interests in and around the canal has made the waterway a flashpoint for US–China competition over spheres of influence, with many in Washington seeing the increased Chinese activity as an invasion of 'America's backyard'. Given the continued trade tensions between the world's two biggest economies, this is not a situation that is likely to be resolved any time soon.

Unlike many other canals which use seawater, the Panama Canal is fed by freshwater from the artificial Alajuela and Gatun lakes. It requires around 200 million litres of water to flow down a series of tiered locks into the sea in order for each ship to pass through and so large amounts of rain are required to keep the system functioning with the locks as much as 26 m above sea level. As a result of severe drought, water levels in Lake Alajuela fell dramatically and restrictions on movement were introduced in July 2023. In 2022, an average of forty ships a day passed through the canal, a number which has now dropped to thirty-two to save water and has had an impact on global trade and the availability and price of goods.

THE STRAIT OF MALACCA

The Strait of Malacca is 580 miles (930 km) in length, lying between the Malay Peninsula and the Indonesian island of Sumatra. It is one of the world's narrowest chokepoints, with its narrowest point being only around 1.5 nautical miles. Despite its size, it is one of Asia's most critical waterways, since it provides a crucial connection between China, India and Southeast Asia. Piracy, which was a real problem in the early part of the twenty-first century, was considerably reduced for a number of years, but by 2019 there were, once again, thirty piracy incidents in the Strait. The annual haze, the result of seasonal bushfires in Sumatra, can reduce visibility to a mere 200 m, a distance much shorter than many of the longest vehicles transiting the area, and represents a major hazard to shipping. One unique historical challenge facing vessels in the area is that there are around thirty-five shipwrecks in the traffic separation scheme (TSS) channel and, in August 2017, the US Navy Destroyer USS *John S. McCain* collided with a merchant ship with the tragic loss of ten of the ship's crew. The Strait's importance has been highlighted as tensions have increased in the South China Sea. In economic terms, 64 per cent of China's maritime trade passes through the waterway while 42 per cent of Japan's trade does the same. China's sweeping and illegal claims to islands in the South China Sea have angered many of their neighbours, including Vietnam, Taiwan, Malaysia, the Philippines and Brunei, while the threat to international freedom of navigation has raised tensions with the United States and its allies. The South China Sea itself must now be seen as one of the great potential global flashpoints of the future.

BAB EL-MANDEB STRAIT

The Bab el-Mandeb Strait, lying between the Horn of Africa and the Middle East, links the Mediterranean Sea with the Indian Ocean via the Suez Canal. It is both a primary waterway for transporting the world's oil and natural gas and a high-risk area for piracy. In May 2020, a UK chemical tanker was attacked off the Yemeni coast, the ninth such attack in the area that year. The amount of crude oil and refined petroleum products transported through the waterway rose to about 4.8 million barrels per day in 2016, from about 3.3 million barrels per day in 2011. China has established its first ever foreign military base in Djibouti, lining it up to be yet another area of contention with the United States in the ongoing battle for global strategic dominance. With two of the world's main maritime

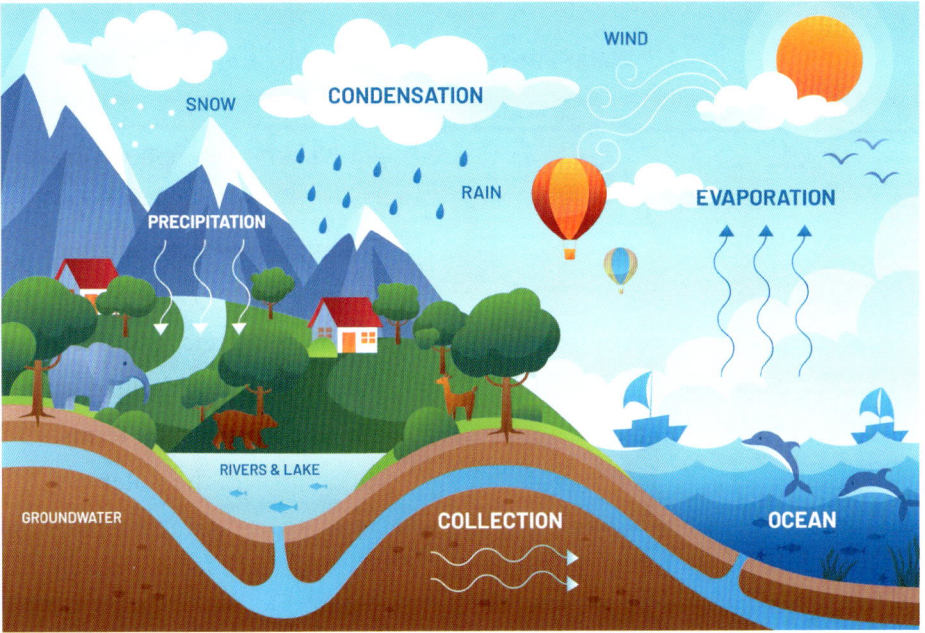

The water cycle. Source: freepik.com

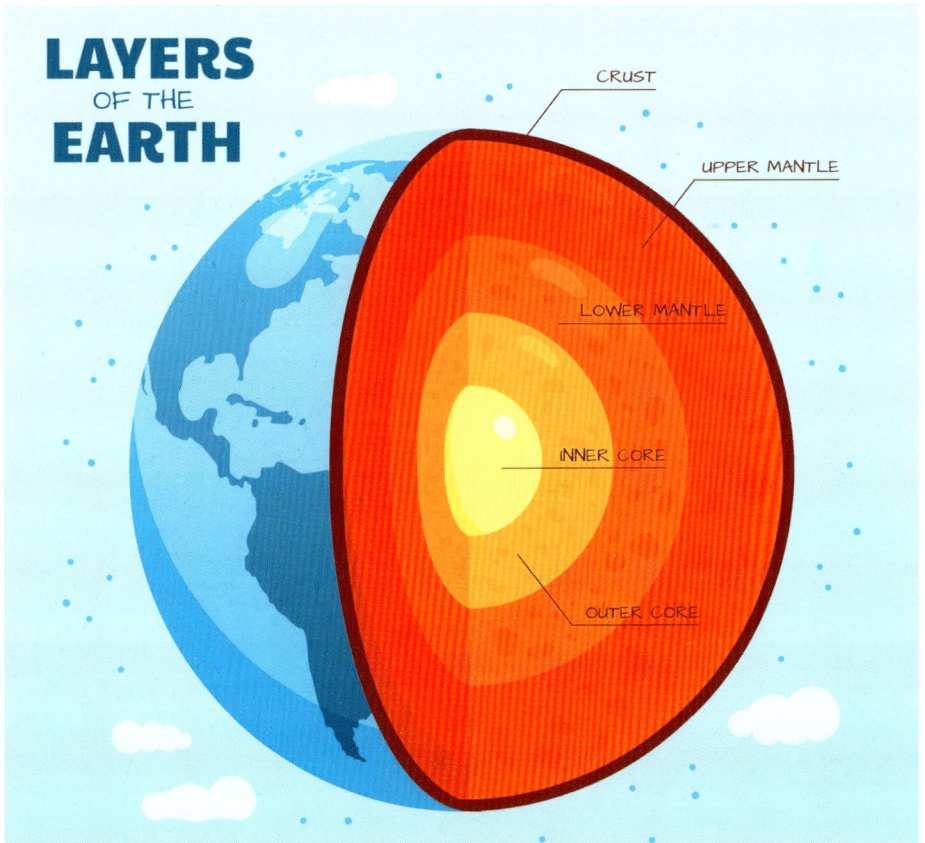

The structure of the earth. Source: freepik.com

Water Stress by Country: 2040

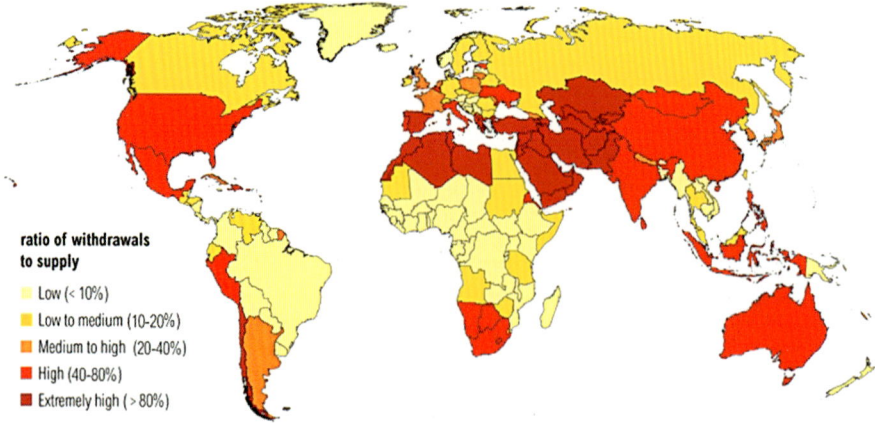

ratio of withdrawals
to supply

- Low (< 10%)
- Low to medium (10-20%)
- Medium to high (20-40%)
- High (40-80%)
- Extremely high (> 80%)

NOTE: Projections are based on a business-as-usual scenario using SSP2 and RCP8.5.

For more: ow.ly/RiWop

WORLD RESOURCES INSTITUTE

Water-stressed states around the world. Source: World Resources Institute, wri.org/aqueduct

Provincial map of Spain. Source: Wikimedia Commons, Emilio Gómez Fernández and Javi C. S.

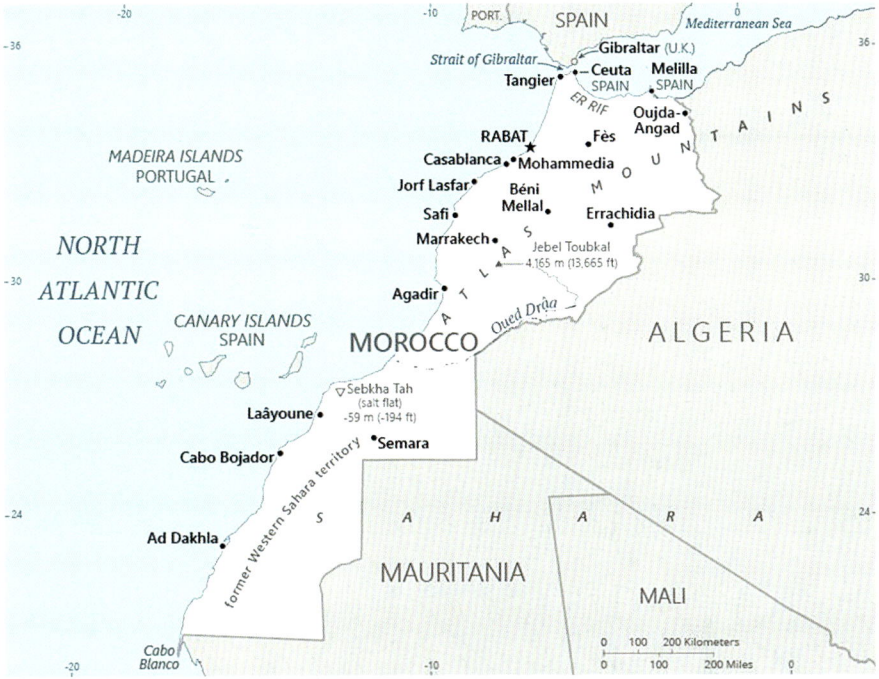

The Kingdom of Morocco. Source: *The World Factbook 2021*, Washington DC: Central Intelligence Agency

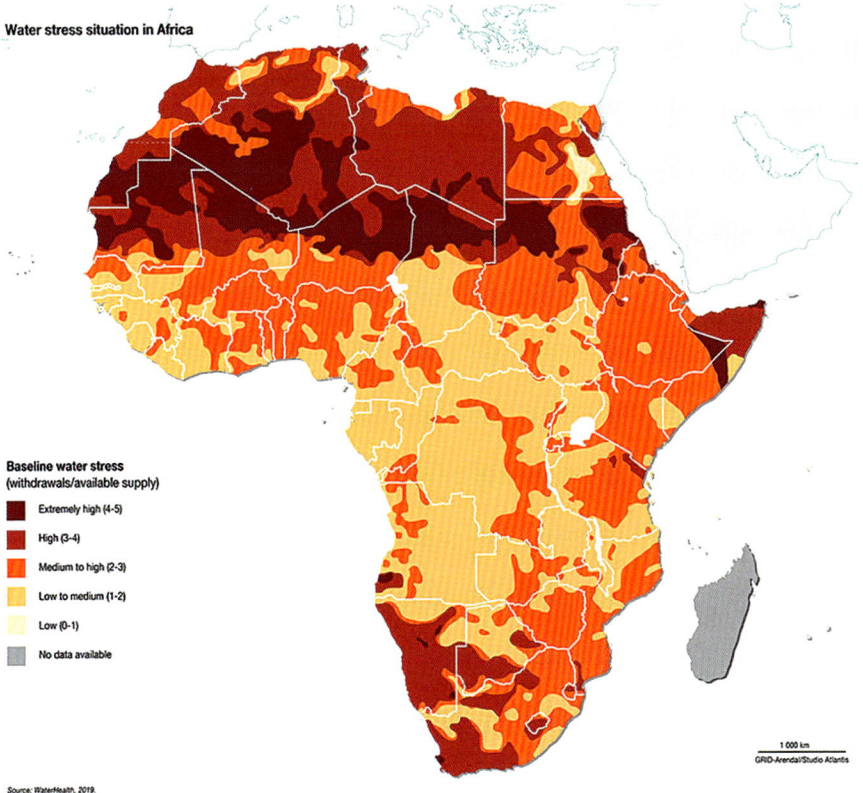

A detailed illustration of the water-stress situation in Africa. Source: GRID-Arendal resources library, grida.no/publications/471

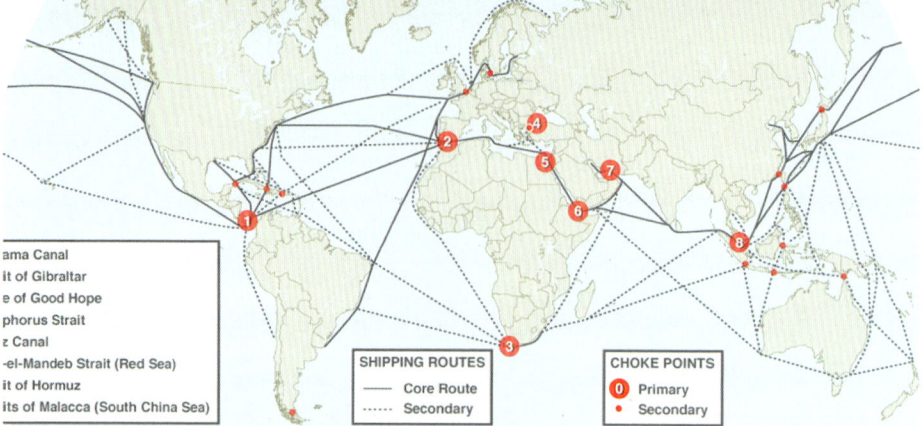

PRIMARY GLOBAL MARITIME CHOKE POINTS

ama Canal
it of Gibraltar
e of Good Hope
phorus Strait
z Canal
-el-Mandeb Strait (Red Sea)
it of Hormuz
its of Malacca (South China Sea)

SHIPPING ROUTES
— Core Route
····· Secondary

CHOKE POINTS
0 Primary
· Secondary

Maritime chokepoints.

Source: *American Journal of Transportation*

Water pump on Broad Street, Soho, London.

Source: Wikimedia Commons, Jamzze

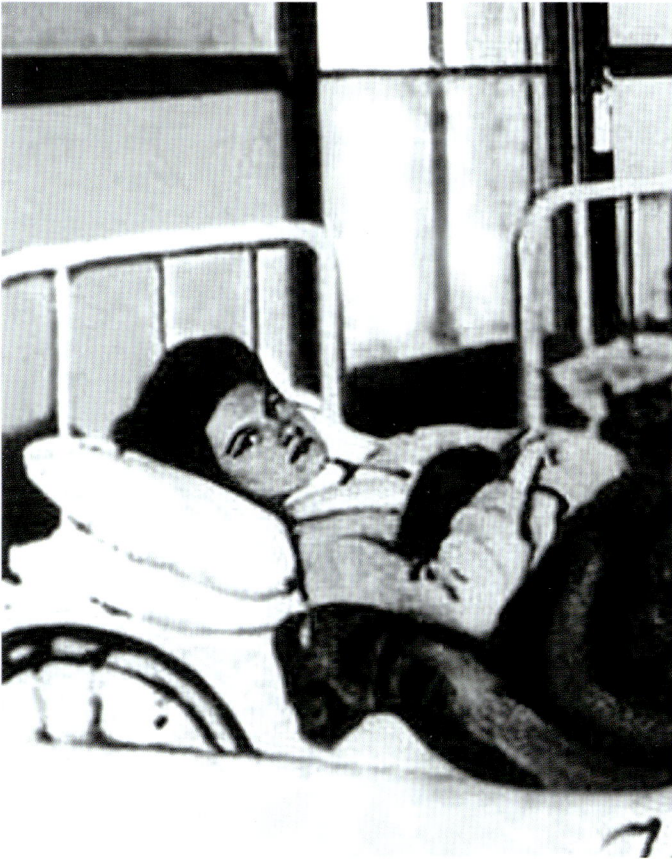

Mary Mallon (Typhoid Mary) in hospital. Source: Wikimedia Commons

Winston Churchill, Franklin D. Roosevelt and Josef Stalin at the Yalta Conference, February 1945.
Source: Wikimedia Commons, US Army Signal Corps

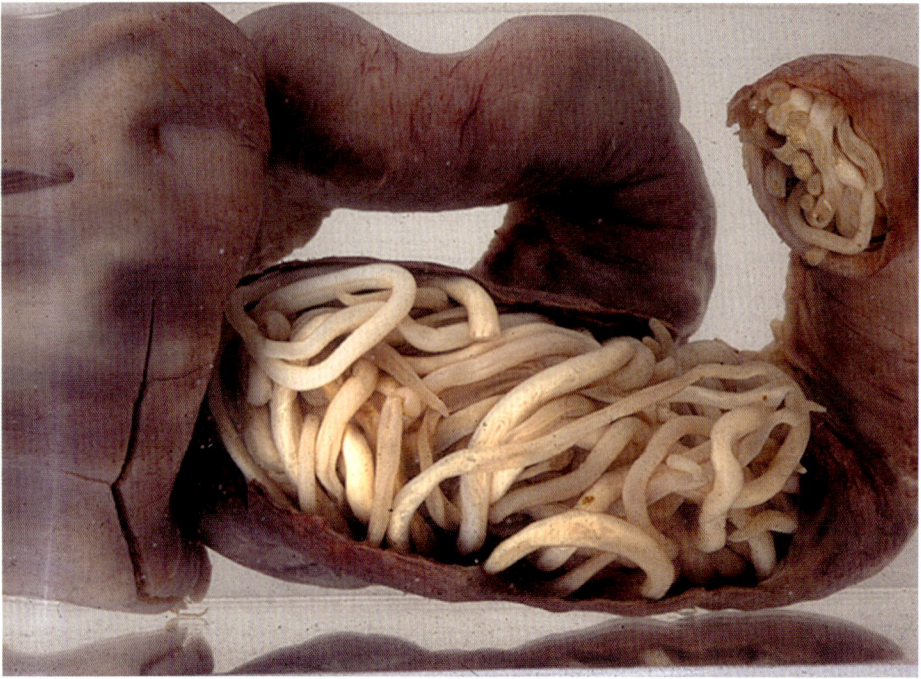

A piece of intestine, blocked by worms and surgically removed from a three-year-old boy at Red Cross War Memorial Children's Hospital. Source: Allen Jefthas, South African Medical Research Council

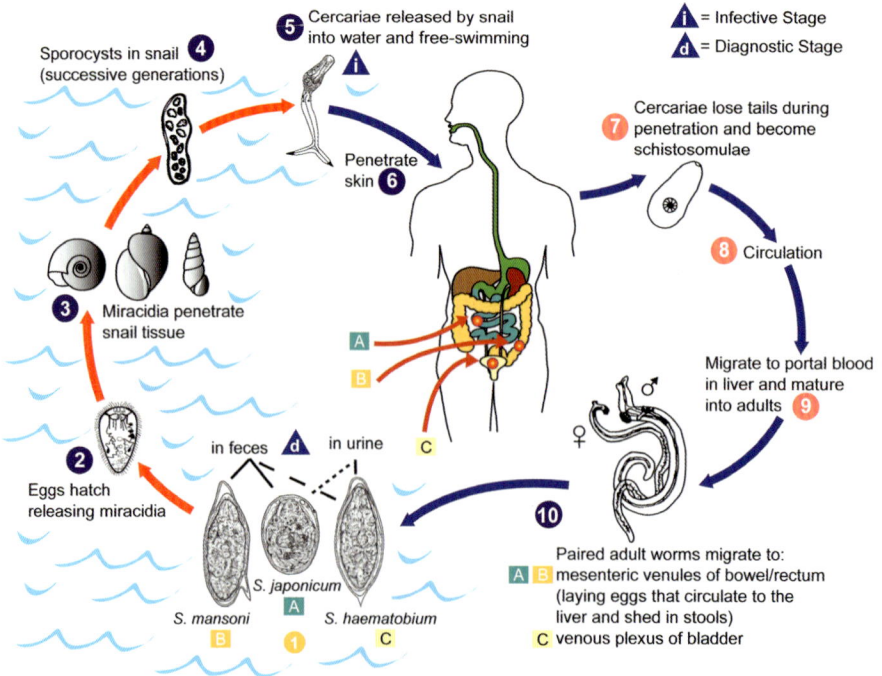

The schistosomiasis life cycle. Source: Wikimedia Commons, CDC

How Thirsty Is Our Food?

Liters of water required to produce one kilogram of the following food products*

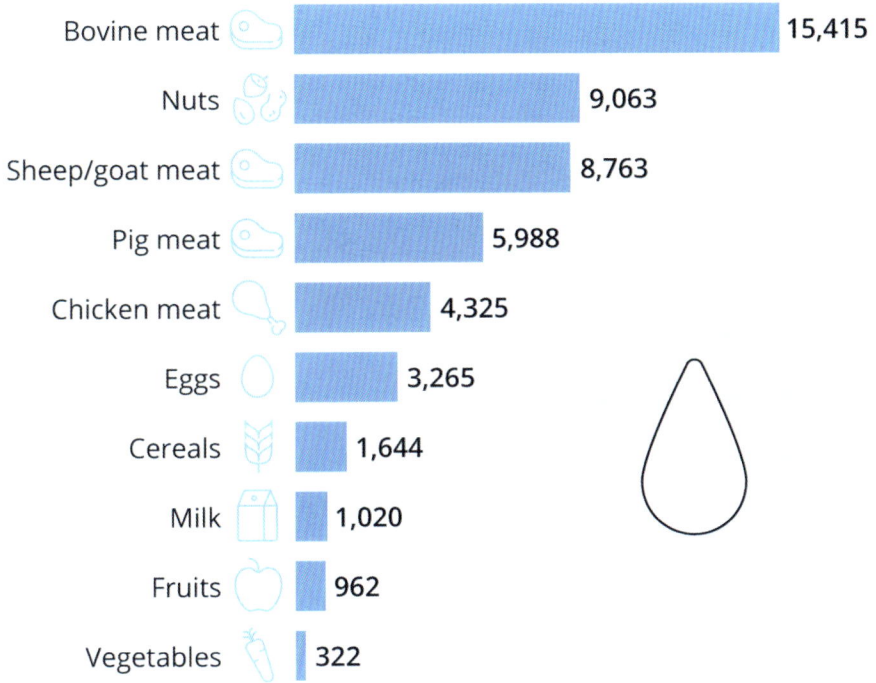

Food	Liters
Bovine meat	15,415
Nuts	9,063
Sheep/goat meat	8,763
Pig meat	5,988
Chicken meat	4,325
Eggs	3,265
Cereals	1,644
Milk	1,020
Fruits	962
Vegetables	322

* Global averages
Source: Water Footprint Network

statista

How thirsty is our food? Source: Statista, statista.com/chart/9483/how-thirsty-is-our-food/

Clevedon Pier. © Clevedon Pier & Heritage Trust Ltd

A beach cleaner holds a handful of freshwater snails pulled from the banks of Lake Victoria in Uganda.

chokepoints at the northern and southern ends of the Red Sea, its geopolitical importance can only increase.

SUMED

Finally, it is worth giving a brief mention, in the context of chokepoints (especially in relation to the Red Sea), to the 200-mile-long Sumed pipeline which carries oil through Egypt from the Red Sea to the Mediterranean. It runs from the Ain Sokhna terminal in the Gulf of Suez to Alexandria. The pipeline, with a total capacity of 2.34 million barrels per day, is the only alternate route by which to transport crude oil from the Red Sea to the Mediterranean Sea if the Suez Canal becomes impassable for any reason.

All the potential risks from global maritime chokepoints, whether related to structure or security, will need to be mitigated for as long as we continue to transport goods and fuel across our oceans and seas.

Even if we succeed in decarbonising western economies, we will still be dependent on strong and functioning economies in other parts of the world as our trading partners. Given the huge dependence that China, for example, has on fossil fuels and the huge length of time it is likely to take to reduce its dependence on Gulf oil, we are going to require long-term global co-operation to ensure the security of maritime transport (as well as pipelines). While the model of international integration of operations in dealing with piracy in Somalia is perhaps encouraging, a great deal more must be done to share the burden in a world whose economic interconnections have brought both new opportunities and new vulnerabilities.

Those nations such as the UK who are not only huge trading nations but, in our case, an island, are economically hugely dependent on the free flow of maritime trade and need to ensure that we invest in the necessary mechanisms capabilities to ensure the security of the high seas. The UK's investment in the Royal Navy, particularly the carrier strike capability, needs to be matched in both funding and intent by our major trading partners if we are to navigate, in every sense, the tricky waters ahead.

PART 3

IN SICKNESS AND IN HEALTH

CHAPTER 13

HEALTH

Humans are watery people. We are composed largely of water and cannot live for long without it. It is essential for good health, and disruption of our natural balance can result in serious illness. The story of water is, therefore, an integral part of the human story from our evolution, to our embryological development, to the functioning of our vital bodily processes.

EVEN BEFORE WE ARE BORN

While still inside our mother's womb, as a foetus, we are surrounded by the fluid contained in the amniotic sac. This amniotic fluid has a vital role in ensuring our safety and health up to birth. At a physical level, it acts as a shock absorber to protect the foetus if there is any trauma to the mother's abdomen and it also provides a cushion between the foetus and the umbilical cord to reduce the risk of compression. The antibacterial properties of amniotic fluid protect the foetus from any infectious agents which may be present, and it acts as a space and source of nutrients that allows the normal development of organs, containing as it does, proteins, vitamins, electrolytes and immunoglobulins from the mother. The fluid is generated from maternal plasma until around week sixteen, when foetal urine also contributes to the fluid volume. It is absorbed through the foetal tissue and skin until around week twenty-five, after which the fluid is mainly absorbed by the foetal gut up until birth. It is our own sort of pre-birth, mini-water cycle. It reaches its maximum level, of around a litre, at thirty-six weeks. Every mum who has experienced her 'waters breaking' will recognise what a large volume it can seem!

A great deal of research is constantly being undertaken to find out what relationships there may be between contaminated water and the outcome of both pregnancy itself and the possible health issues that may not manifest themselves until later in childhood. We know, for example, that exposure to lead in drinking water during pregnancy, which usually comes from water passing through rusted pipes, can have an impact on babies' intelligence quotient (IQ) and increases attention-related behavioural problems later in life. There is some evidence that certain pesticides, which can seep into the water supply because of runoff from crop production, can increase the risk of reduced foetal growth. Arsenic, which can reach water through natural deposits or industrial pollution has also been linked with increased chances of miscarriage, pre-term birth and birth defects. More recently, there has been an increased focus on a potential link between the presence of disinfection byproducts (DBPs), which are formed when chlorine and bromine interact with natural organic materials in water, and an increased risk of birth defects. Since the health of a child at birth is an important predictor of things like future educational attainment and earning potential as an adult, it all highlights the need to ensure a clean water supply for everyone, a subject that we will deal with shortly.

OUR KEY COMPONENT

Every system that our body contains, and every function that it carries out, is dependent to one extent or another on water. It is the building block for new cells on which every tissue depends. It is key to transporting and metabolising the nutrients we require for our energy and good health. It protects sensitive tissues such as our eyes from harm and it is the key ingredient in our saliva.

It helps us regulate our body temperature through respiration and sweat and it helps us flush the waste products out of our bodies. That's not a bad range of functions for such a simple chemical compound.

Water makes up around 60 to 70 per cent of the human body, ranging from around 75 per cent of body mass in infants to 50 to 60 per cent in normal adults, down to as little as 45 per cent in the elderly. These changes at different ages are associated with the different proportions of water in our bodies' basic components – muscle, fat, bone and other tissues – at different stages of our lives. Our brains and kidneys have the highest proportion of water at around 80 to 85 per cent. Our hearts and lungs come next at between 75 to 80 per cent water with our liver, skin and muscles at 70 to 75 per cent. Perhaps surprisingly, the water content of blood is around 50 per cent (although plasma, the liquid portion of the blood, is about 90 per cent water, carrying blood cells, nutrients and hormones throughout the body). While bones have 20 to 25 per cent water, fat has only 10 per cent and teeth a mere 8 to 10 per cent.

THE IMPORTANCE OF WATER BALANCE

We can think about how water is contained within our bodies as a series of fluid compartments. Anatomically, these are not literal compartments, but they represent a functional separation of water, salts and suspended elements.

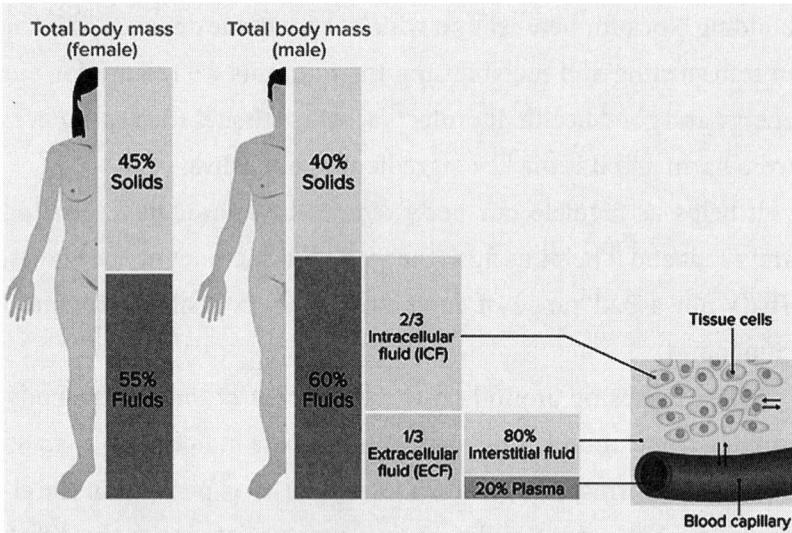

Intracellular fluid (ICF) is the water that lies inside our cells themselves and this constitutes around 60 per cent of all our bodily water. This would be around seven gallons (25 L) in an average adult male.

Extracellular fluid (ECF), as the name implies, surrounds all the cells in the body and makes up another third of the body's water. It has two main components, plasma (the fluid component of the blood) and interstitial fluid (IF) which surrounds all the cells that are not in the blood. Other, smaller, components of ECF include the cerebral spinal fluid surrounding our brain and spinal cord, the synovial fluid that provides the lubrication in our joints and the fluid in our lung cavities and cardiac sac. Although the fluid in our bone and connective tissue makes up about 15 per cent of the total amount of water in our bodies, it has little importance in fluid balance because it moves at such a slow rate.

Water and electrolytes (salts) are constantly moving in tiny amounts to maintain our body's equilibrium process, which is called homeostasis.

The body's aim is to keep the concentration of intracellular and extracellular fluid roughly the same. If the concentration of the extracellular fluid rises too much in comparison to the intracellular fluid, then water will be drawn out of the cells, damaging their function. Conversely, if the concentration of the extracellular fluid falls in relation to the intracellular fluid, then water will be drawn into the cells and they may burst. Sensitive receptors in the body can detect small changes in volume and concentration (mainly of sodium) and keep the system in equilibrium. Despite the huge variation in both our water and sodium intake (common salt is sodium chloride), balance is maintained within very narrow limits.

An excess or a deficit of just a few hundred millilitres is enough to trigger sensitive physiological mechanisms to redress the imbalance. If we become dehydrated, the plasma volume will fall and the fluid concentration will increase. This triggers the release of a substance called antidiuretic hormone (ADH) from the brain which instructs the kidneys to increase water reabsorption. This makes us produce a lower volume of urine with a higher concentration, thus conserving water. The reverse process occurs if there is an excess of water. ADH release becomes inhibited and more diluted urine is produced.

So, there are effectively two drivers of the behavioural change that we know as thirst. The first relates to the volume of blood. If we lose too much blood volume, for example by bleeding, vomiting or diarrhoea, our blood pressure will fall and we will get what is known as 'hypovolemic driven thirst'. A combination of physiological and behavioural pathways are used to correct the problem.

The second driver is when the concentration of the extracellular fluid increases, drawing water out of the cells. This can occur due to a high intake of dietary sodium or the loss of a lot of water through perspiration, respiration, defecation and urination. This is called 'osmometric thirst' and will result in chemical responses that will help the kidneys to retain water and sodium, as well as stimulating our urge to drink more water. All this is co-ordinated in a small part of our brain called the median preoptic nucleus which also synchronises our sleep pattern and temperature regulation. Destruction of this part of our brain results in the reduction, or even complete loss, of the desire to drink, with potentially fatal salt concentrations accumulating in the extracellular fluids.

THE IMPORTANCE OF SALT

Salt (and therefore sodium) consumption in humans has varied over time, affected by social and dietary habits. As hunter-gatherers we would have eaten very little salt and because of this it is thought that our kidneys have evolved to conserve nearly all of our sodium. Having salt as a preservative for food (which really seems to have begun with the Egyptians) would clearly have increased dietary intake until the advent of refrigeration, which was not really developed until the early 1800s.

Today, there is increased public awareness about the amount of salt contained in processed foods and the impact that this can have on human health. It is estimated that around half of the American population today is hypertensive, salt sensitive or both. Salt sensitivity, the temporary rise in blood pressure that is associated with a rapid ingestion of salt, affects about 25 per cent of the population in the US, but only half of those who have salt sensitivity have actual hypertension, so the relationship is not entirely clear-cut.

What we do know is that too much dietary salt can, at least in some individuals and over time, raise blood pressure and that hypertension (high blood pressure) is a major risk factor for heart disease and stroke. We also know that a reduction in dietary salt will not only reduce the incidence of high blood pressure but will also reduce the risk of death and illness from adjacent cardiovascular diseases. A modest reduction in salt intake, maintained over a period of time, produces a fall in blood pressure in both hypertensive patients and normotensive patients (those whose blood pressure is within the normal range), irrespective of their sex or ethnic group. It is, therefore, worth everybody doing so.

Interestingly, at the other end of the scale, while most animals will respond to periods of salt deficiency by ingesting it directly, humans are different. Animals may go to salt licks, places where they can get essential mineral nutrients from natural salt deposits and man-made salt licks might attract rabbits, foxes, squirrels and birds. Humans, however, choose to eat salt only in the form of food and reject the taste of very salty solutions. This is something we seem to know almost instinctively, as everyone knows the trick of drinking salty water to induce vomiting if a person has ingested something they shouldn't have. Humans are also unusual in that we will only ingest the chloride form of sodium while most other animals will ingest any other sodium compound.

What happens when our finely tuned system to balance our water and sodium levels goes wrong? It can happen if we have too little water, too much water or water in the wrong place.

THIRST

How thirst – the basic instinct that makes all animals drink – is switched on and off is the result of a complex interaction of both the physiological and behavioural processes we discussed above. How much fluid we drink in a day is dependent not only on our bodies' need for water but also on the taste, availability and appeal of drinks, as well as social and environmental stimuli. Pubs and bars would not make much money if their only function was to simply maintain fluid balance, though we will return to the subject of alcohol intake later. The early symptoms of thirst – a dry mouth, throat and lips can progress to tiredness, irritability, dizziness and headache. Of the two drivers to thirst that we have already looked at, hypovolemic (reductions in extracellular fluid volume) and osmometric (increases in extracellular concentration), the hypovolemic thirst mechanism is less sensitive and only kicks in when the extracellular fluid falls by about 10 per cent.

Thirst is a lag indicator as it only occurs when around 2 per cent of body weight is already lost. This means that it is a poor management tool for our body's hydration. By the time you feel thirsty you may already be experiencing symptoms such as stress, agitation and forgetfulness, which is why it is important to drink plenty of water to keep thirst at bay. The human body can lose 2.5 to 5 pints (1.5–3 L) every hour due to sweat when exercising in a hot environment on top of the moisture lost through breathing. Exhaled air has more water than inhaled air, as sugar is formed in respiration, along with carbon dioxide. About 400 ml of water is lost each day by respiration alone. Whatever the cause of dehydration, as our bodies react to conserve water, our cardiovascular system must work harder to keep our blood pressure up, so urine volumes fall sharply. At around 4 per cent of body water loss this correction mechanism becomes more difficult, blood pressure can fall and fainting and dizziness can occur. At 7 per cent total body water loss, organ damage starts to occur because of diminished blood supply.

In severe cases, seizures can occur as a result of imbalances in the amount of sodium, potassium and other electrolytes.

The reduction in sweating can also lead to another serious problem – the increased risk of overheating. The most serious form of this is heatstroke, which can happen when body temperature reaches 40°C (104°F). As the sweating mechanism fails, the body is unable to cool itself down and body temperature may rise to 41°C or more within as little as ten minutes. This can be fatal and is one of the reasons behind the fact that death rates tend to rise more sharply during periods of extremely hot weather than during periods of extremely cold weather. Those most at risk of heat injury include the elderly, especially those over seventy-five, babies and young children, those with respiratory or cardiac problems, those who misuse alcohol or drugs, those already suffering from infection and those who are physically active, such as athletes and manual workers.

Some chronic health problems that result in a prolonged increase in urination, such as diabetes and kidney disease, can also lead to dehydration, which itself can contribute to an increased risk of urinary tract infections and kidney stones, as well as the wider range of symptoms discussed above.

TOO MUCH, TOO LITTLE, WRONG PLACE

How long human beings can live without water has long been a subject of debate, but there is a consensus that three to five days is the maximum that we can survive with no source of water at all.

In April 1979, however, an eighteen-year-old bricklayer called Andreas Mihavecz from Bregenz in Austria was taken into police custody by mistake following a car accident. Unfortunately for him, each of the policemen who were supposed to supervise him either forgot about him or thought that the others had already released him. As a result, poor Andreas was left in his cell for eighteen days and only survived by ingesting condensed water from the walls, losing 53 pounds (43 kg) in weight before the appalling mistake was discovered and he was freed. Amazingly, he was awarded only 250,000 Austrian schillings (around $23,000) in compensation although he enjoyed the minor reward of making it to *The Guinness Book of Records*.

Of course, just as too little water can be deadly, so can too much. Water intoxication is a condition which arises when the levels of sodium, potassium and other electrolytes become too depleted and cause potentially fatal disturbances in brain function.

This has become a high-profile issue in recent years because of the use of the drug MDMA (Ecstasy) particularly by young people in nightclubs. Ecstasy has two simultaneously damaging effects. It causes low sodium levels to occur because inappropriate secretion of antidiuretic hormone (ADH) combines with excessive fluid ingestion due to inappropriately increased thirst and creates a double diluting effect. In more severe cases, initial vomiting and disturbed behaviour can be followed by seizures and even death. In milder cases, drowsiness and disorientation have been observed in users for up to three days.

Young children (particularly those under nine months) are especially vulnerable to water toxicity as it is easy for them to take in a large amount of water relative to their small body mass and the total sodium stores in their bodies.

Excessive water intake can also occur in certain psychiatric states. Primary polydipsia (PPD) is a condition that results in the patient drinking an excessive amount of fluid in the absence of the normal physiological stimuli which cause us to drink. It tends to be found most often in schizophrenia, but it also occurs in anxiety disorders and, more rarely, conditions such as anorexia nervosa and personality disorders. Some studies have suggested that PPD occurs in 6 to 20 per cent of hospital psychiatric patients.

A more commonly encountered problem occurs among long-distance runners who might be susceptible to water intoxication if they consume too much while they are running because of their sodium levels dropping too low. Drinking isotonic fluids has become the main way in which this problem can be avoided.

Even if we have the correct amount of water in our bodies, clinical difficulties can still arise if it is in the wrong place.

Oedema is the accumulation of excess water in the tissues, most frequently in the extremities. It can be caused by water leaking from blood vessels, a range of medical conditions, including pregnancy, certain drugs, injury or allergy.

If a person's limb is swollen and a finger is pressed into it for a few moments and then removed again and it leaves an indentation behind, it is a phenomenon known as 'pitting oedema'. This is an important clinical and diagnostic sign. If swelling has been present for more than three months it is 'chronic' oedema which can lead to complications such as infection, fluid leakage, ulceration or cellulitis.

The lymphatic system is a kind of drainage system throughout the body that helps to fight infection and remove excess fluid. When this system is not working properly it can produce a type of oedema known as lymphoedema. There are two types of lymphoedema: primary and secondary. Primary lymphoedema, which is genetically linked and occurs due to faulty development of the lymphatic system tends to occur in infancy, adolescence or early adult life. Secondary lymphoedema can be the result of cancer treatment, infection, injury or lack of movement in the limbs. While the swelling may be intermittent at first, without treatment it can become severe and persistent. Symptoms include a heavy aching feeling, repeated skin infections, hard or tight skin with folds developing in the skin and with fluid sometimes leaking through it. It is thought that lymphoedema affects more than 200,000 people in the UK and early treatment is essential if the condition is not to become serious and persistent. It not only causes pain and discomfort for patients but can be a major cause of social isolation and, thankfully, both health services and community groups are beginning to understand the complex needs of those who suffer from this condition.

There are two potentially life-threatening forms of oedema which are worth mentioning – when it occurs in the lungs (pulmonary oedema) and when it occurs in the brain (cerebral oedema).

Pulmonary oedema occurs when fluid accumulates in the small air sacs of the lungs, making it difficult for oxygen and carbon dioxide to be exchanged and resulting in difficulty breathing and sometimes in chest pain. It occurs most commonly in people with heart or kidney failure. In cases of heart failure, the oedema results from the failure of the pumping action of the heart itself which, because circulation is a closed system, results in a 'backing up' of blood. The resulting pressure causes fluid to be forced out of the small blood vessels in the lungs and into the air sacs themselves. The oedema is usually treated with drugs that try to increase the heart's ability to contract effectively (a better pump) or by diuretics which lower the total volume of blood in the system (less in the pipes).

Cerebral oedema (or swelling of the brain) is the abnormal accumulation of fluid within that organ and causes the disruption of normal nerve function and results in increased pressure inside the skull (raised intracranial pressure). It can occur in conditions such as stroke, haemorrhage (including during traumatic injuries) or brain cancer. Symptoms can include headache, nausea, vomiting, drowsiness and seizures, with serious cases sometimes resulting in coma and death.

The reason that cerebral oedema is so dangerous is that it relates to our anatomy. Since the brain is contained within the bony confines of the skull (along with the blood vessels and the cerebral spinal fluid that bathes the brain) any increase in pressure may cause even more damage, due to compression of the brain tissue and blood vessels, than the primary cause of injury.

Where intracranial pressure continues to rise, the brain may be pushed down through the base of the skull into the spinal canal which can damage the brain stem. Since this is where the respiratory system is controlled, breathing can be interrupted, with fatal consequences.

Cerebral oedema can also be caused by low sodium levels and is a complication associated with altitude sickness. Headache, nausea, vomiting and drowsiness are symptoms that must be taken very seriously at altitude as they can be early warning signs of developing cerebral oedema in otherwise healthy individuals. While attending a medical conference in Denver, my wife and a group of fellow doctors were lucky to have a potentially deadly case of cerebral oedema in their party identified by someone who quickly recognised the symptoms from experience – the hotel lift attendant!

WHAT CAN WE DO TO HELP OURSELVES?

There is no clear consensus about exactly how much water we should drink each day, largely because, on top of our basic requirements, our needs are dependent on several external factors. Climate is clearly one of these considerations, with heat and humidity requiring extra fluid intake. Increased activity such as sports or outdoor labour will exacerbate this trend, as will pregnancy or breastfeeding. Any co-existing health problems, such as diabetes or any infection that induces a fever will also require us to take more water on board.

One recommendation that is frequently cited is the eight-by-eight rule, which refers to drinking eight 8-ounce glasses of water a day, resulting in an intake of around 4 pints (2 L). The US National Academies of Sciences, Engineering, and Medicine suggests 11.5 cups (2.7 L) a day for women and 15.5 cups (3.7 L) a day for men.

Diet is also an important consideration in determining our fluid needs as tea, coffee and popular soft drinks containing caffeine (which is a diuretic and will cause more water to be lost through urine) will increase the amount of water we need to drink to stay in balance. Salty and spicy foods also require extra dietary water, something that may exacerbate our thirst if we are eating salty and spicy snacks at the same time as we are drinking alcohol.

As we have discussed earlier, thirst is a poor indicator of hydration and because it is a lag indicator there is more than a little truth in the old saying that 'if you are thirsty, you are already dehydrated'. While thirst might not be a good guide to dehydration, we are all naturally equipped with another ready reckoner. One of the mechanisms that we can deploy to determine our state of hydration (or dehydration) is to understand what our urine is telling us. The average person will pass around 1.5 L of urine per day and it will naturally change colour at different times. In the morning, urine should be a pale straw colour if there was sufficient water intake the previous day. If we drink enough water then our urine should become clearer as the day goes on, so if it becomes darker, then there is a shortfall. One of the instructions given to our troops in warm climates is to 'pee clear once a day', a simple method of determining whether they are drinking enough.

As we get older (and that means any time over fifty years), our thirst mechanism starts to diminish and so it is very important to ensure that older people are maintaining their levels of hydration, particularly in spells of warm weather. While the death rate increases during very cold spells, it tends to rise even further in very warm spells where a contributory factor is usually that older and vulnerable people are failing to take sufficient liquid on board. Checking the intake of elderly relatives or neighbours in hot weather can be, literally, a lifesaver.

FOOD AND ALCOHOL

Since we get around 20 per cent of our daily water from the food we eat, it is worth knowing which foods contain higher levels as the more 'wet' foods we eat, the less we will depend on our drinking water to top us up.

The water content of foods is highly variable. Foods such as strawberries, tomatoes, lettuce, spinach or celery contain 90 to 99 per cent water. Apples, oranges, pineapple, grapes and yoghurt are in the 80 to 89 per cent range. Prawns (shrimp), bananas, cottage cheese or avocados will be 70 to 79 per cent while chicken breast, salmon or pasta will provide 60 to 69 per cent. Steak, hotdogs and ground beef contain 50 to 59 per cent water, pizza will be down at around 40 to 49 per cent while bread and cheddar cheese will be even less at 30 to 39 per cent. For those with a sweet tooth, cakes and biscuits will only provide 20 to 29 per cent water while at the bottom of the league, foods such as raisins or butter fall under 20 per cent.

So, to stay hydrated, keep as cool as possible, drink plenty of water, avoid salty and spicy foods and ensure that the foods we do eat are high in water content.

It would be impossible not to discuss one topic that is often on our minds when we think of hydration – alcohol consumption. Alcohol is a diuretic which means that it will cause us to pass more urine whatever our overall state of hydration. This is because it reduces the production of antidiuretic hormone (ADH) so more water is lost through urination. Every standard unit of alcohol equals 10ml or 8 grams (g) of pure alcohol, roughly the amount an average adult can process in an hour. For every 10 milligrams (mg) of alcohol that we ingest we will emit around 100 ml in urine more than we drink. So, the higher the alcohol content of a drink, the more urine we will lose and the more dehydrated we will become. So, choose your drinks carefully. Beer contains around 4 to 6 per cent alcohol, wine, 12.5 to 14.5 per cent, with vodka and rum containing around 40 per cent. Going further up the scale, gin contains 37.5 to 50 per cent alcohol while whisky is in the 40 to 60 per cent range.

Finally, the dreaded subject of the hangover. This usually begins when our blood alcohol content drops significantly and is at, or near, zero. Typical symptoms include fatigue, headache, thirst, nausea, sweating, irritability and increased sensitivity to light or sound. Hangovers are primarily generated by dehydration due to the diuretic effect of alcohol. But alcohol also contributes to the awful state in several other ways. It irritates the stomach lining, reduces the production of stomach acid and delays the stomach emptying which can cause the symptoms of nausea and vomiting. Inflammatory responses triggered by the immune system can also result in diminished concentration, memory and appetite. Irritation of the blood vessels can produce headaches and the fall in blood sugar that results from alcohol consumption can lead to weakness, shakiness and changes in mood. To top it all off, even if you manage to get some sleep after a heavy night out, alcohol will diminish the quality of the sleep that you do get and leave you more tired than usual in the morning.

It is not alcohol alone, however, that produces hangovers. Substances called congeners are chemical by-products of the fermentation process that give alcoholic drinks their distinctive flavour. Such compounds include methanol, tannins and acetaldehyde. These are found in larger quantities in dark drinks like red wine or brandy in comparison to clear drinks such as vodka or gin (vodka, in fact, has almost no congeners at all). Those who find themselves most susceptible to hangovers should probably avoid dark drinks where possible.

The key to defeating a hangover lies in hydration, so drinking plenty of water before bedtime is a good idea, as is keeping a large glass (or bottle) of water by the bedside in case you wake up during the night. One good rule is to alternate between alcoholic drinks and water (or soft drinks) which will both reduce the amount of alcohol that you consume over time and minimise dehydration. If you can face it in the morning, eating a good breakfast will help to keep your sugar levels up and provide important vitamins and minerals, all of which will contribute to reducing hangover symptoms. If you cannot face a hearty breakfast, then bananas and bread will at least help.

Antacids may assist in dealing with the gastric consequences of a hangover and going back to bed to make up some of the sleep deficit will help, which makes it a bad idea to drink midweek! How quickly each of us metabolise alcohol is extremely variable so learning about your own alcohol tolerance is important.

Drinking moderately, drinking slowly and drinking lots of water is the key to a happier morning after. I used to tell my patients, however, that happiness is a key part of health, and an enjoyable night out can be a real tonic and stress reliever. I used to tell them, 'Everything in moderation – including moderation'. It was probably not the best piece of advice I ever gave them, but it was certainly one of the most popular!

CHAPTER 14

SICKNESS

Nature is not always our friend. While water is essential for life and good health, it can also be a vehicle for death and disease.

It is estimated that as many as 3.5 million people die from waterborne disease each year, with around 2.2 million of these being children – and this is in the twenty-first century.

These numbers are so huge that it can be difficult to really grasp their significance. In terms that are easier to comprehend, the global deaths from waterborne diseases among adults are equivalent to a death toll of three 9/11 atrocities a day or six *Titanic* sinkings a day and the number of dead would fill Wembley Stadium's 90,000 seats every ten days. What is truly shocking, however, is that the number of deaths among children is the equivalent of eighteen Boeing 777 jets, full of children, crashing every day.

Of course, death rates are only part of the story. Non-fatal infections from waterborne diseases are widespread among those populations who live in parts of the world with inadequate clean water and sanitation. Of the 8 billion people on our planet, it is estimated that 2 billion do not have basic sanitation, such as toilets or latrines and that almost 700 million are forced to defecate in street gutters, behind bushes or into open bodies of water, with all the implications that brings for the further spread of disease. In addition, at least 10 per cent of all food consumed is believed to be irrigated by waste-water, which adds to the risk of disease and economic hardship as sick populations struggle to be productive. Incidentally, again for comparison, the 2 billion people without adequate access to clean water or sanitation is twice the size of the combined populations of the United States, the United Kingdom, Canada, the entire European Union and Japan.

ENDEMIC, EPIDEMIC, PANDEMIC

When we talk about infectious diseases, we often use the terms en-demic, epidemic and pandemic, but what is the difference between them? The first thing to note is that these terms do not indicate the severity of a disease, but simply its prevalence, i.e., the frequency of a disease at any point in time.

A disease is endemic if it is continually present within a specific population or region and the prevalence of the disease is relatively consistent over time. Cholera, for example, is regarded as being en-demic in India.

An epidemic occurs when a disease spreads quickly or unexpect-edly through a particular population, either because an endemic disease suddenly becomes much more prevalent, or if a new disease starts to affect a particular population or region. Ebola virus, for example, became an epidemic in west Africa in 2013 and followed a different pattern from the small and limited outbreaks that had occurred sporadically over the previous forty years.

A pandemic occurs when we see a disease spreading across coun-tries or continents at a rapid rate. Covid-19, which was first detected in late 2019, went on to cause vast global disruption before vaccine rates and public health measures brought it under control. Some of the most devastating pandemics in history include the 1918 'Spanish flu' pandemic, which occurred after the First World War and resulted in around 50 million deaths worldwide, and the greatest pandemic of all, bubonic plague or 'the Black Death', which is estimated to have killed up to a third of the European population in the Middle Ages.

An example of a disease which exhibits all three forms – endemic, epidemic and pandemic – is one of the biggest waterborne killers: cholera.

CHOLERA

Cholera appeared in European medical literature in the mid-seventeenth century and is thought to have originated in the Indian subcontinent where it is an endemic disease today. In 2021, the World Health Organization (WHO) estimated that there were between 1.3 to 4 million cases of cholera, occurring across all continents, worldwide, with seventeen countries in Africa, twelve in Asia, four in Europe and four in the Americas. The worst affected were Haiti, the Democratic Republic of Congo, Yemen, Peru, Somalia, Tanzania and Vietnam who accounted for 80 per cent of all cases. The hardest-hit countries demonstrate the increased vulnerability of those who already suffer from a high underlying risk and who are then subjected to the increased stresses of natural disaster or human-induced conflict. Following the 2010 earthquake in Haiti and the Dominican Republic, for example, the breakdown in basic water infrastructure, including the lack of portable drinking water, was a major factor in causing a devastating cholera outbreak. Following Hurricane Matthew, in May 2016, there was a further upsurge of cholera in Haiti, and in 2017–18, war-torn Yemen faced the largest outbreak of cholera in its history. In 2022, global cholera cases more than doubled and they continued to grow in 2023. Seven countries – Afghanistan, Cameroon, DRC, Malawi, Nigeria, Somalia and Syria experienced outbreaks affecting 10,000 people or more.

LESSONS FROM INDIA

UNICEF estimates that today, half of the world's human excrement remains untreated by proper sewage systems. In 2014, the WHO/UNICEF produced a report that estimated that India accounted for 60 per cent of the global population engaged in open defecation. While this sounds like an appalling number, it is worth appreciating the efforts that India's government has made to try and deal with this crisis, in a domestic population of over 1.4 billion people. As far back as 2000, the country initiated major studies by the National Institute of Cholera and Enteric Diseases (NICED) to determine exactly how widespread the disease had become. A combination of historical data and laboratory typing of cholera strains showed that out of the twenty-eight Indian states that provided strains, eleven were endemic for the disease, suggesting that cholera was endemic over a much wider area of the country than had previously been thought. The NICED study showed that there had been a total of more than 200,000 cases of cholera from 1996 to 2007, which is a considerably higher number than the 37,000 cases India reported to the WHO over the same period. Unsurprisingly, those areas that were determined as high-risk zones lacked proper drinking water and drainage, along with access to basic healthcare.

Historically, India has been strongly associated with global outbreaks of disease. Over the past two centuries, seven cholera pandemics have occurred, with the first originating in India in 1817. An estimated 15 million people are thought to have died in the first three pandemics in the country between 1817 and 1860, with another

23 million dying between 1865 and 1917 during the next three. Arising in the Bengal region of India, near Calcutta, the first pandemic ran for seven years before spreading to southeast Asia, the Middle East, Europe, and east Africa. The movement of British Army and Navy personnel is thought to have contributed to the range and spread of the disease, with India being a key hub of the British Empire at that time.

Cholera went on to become a widespread disease during the nineteenth century and has since killed tens of millions of people. In the second pandemic, over 150,000 Americans died and consequently, cholera became the first reportable disease in the United States.

LESSONS FROM JOHN SNOW

In 1832, a young English doctor, John Snow had first experienced a cholera epidemic in the mining village of Killingworth in Northumberland, in the north of England. After moving to London, he set up a practice in Frith Street, Soho and later became a member of both the Royal College of Surgeons and the Royal College of Physicians. He is remembered as an expert in the development of chloroform as a safe anaesthetic, including during childbirth. In April 1853, Queen Victoria asked Snow to administer chloroform during the delivery of her eighth child, Leopold. She was so delighted by its effect that she asked him to repeat the procedure in 1857 for the birth of her daughter, Beatrice.

When it came to the subject of cholera, Snow had long been sceptical about the widely accepted 'miasma' theory of disease, namely, that it was a result of foul air emanating from rotting organic matter. Snow was a keen observer of patterns of disease, carefully noting and recording observed data. At the time, most Londoners obtained their drinking water from a town pump near to where they lived, and Snow gradually became convinced that there was a link between water supply and a cholera epidemic that hit Soho in 1854. Such was his influence (a year after administering chloroform to Queen Victoria) that he persuaded the local authority to remove the handle of the water pump in Broad Street, Soho, and the outbreak came to an end. The John Snow memorial and pump can be seen in what is now Broadwick Street, five minutes' walk from my own flat.

Perhaps most importantly for modern epidemiology, Snow used a dot map to show clusters of the disease and was able to demonstrate that those Londoners receiving their water from the polluted River Thames experienced fourteen times the rate of cholera compared to those who received their water from cleaner sources further upriver. These observations and the public health implications that flowed from them have had a profound effect on how we have viewed diseases, including waterborne diseases, ever since.

It was not until 1883 that the German physician and microbiologist Robert Koch identified vibrio cholerae with a microscope as the bacterium causing the disease. Having discovered the individual causative agents of tuberculosis and anthrax, as well as cholera, he is widely regarded as the father of modern bacteriology. Moreover, Koch used his discoveries to establish the idea that individual germs caused specific diseases and, by this simple linkage, he created the foundations for modern ideas of public health management.

CAUSES

Cholera is caused by the bacterium vibrio cholerae infecting the small intestine. Most bacteria do not survive the acidic conditions of the human stomach once they have been consumed but those that do survive and reach the small intestine propel themselves through the thick mucus using their tail-like 'flagellae' to reach the intestinal walls where they can attach themselves.

Once attached, they release cholera toxin which is eventually incorporated into the intestinal cells and alters fluid balance, resulting in the loss of a fluid that is rich in electrolytes and producing the diarrhoea that is characteristic of the disease. The number of bacteria required to cause disease varies hugely depending on the bacterial strain and the health of the host, but typically, around 100 million bacteria need to be ingested to cause clinical infection in a healthy adult. Individuals with blood group O seem to be more likely to develop severe cholera than other blood types. The dose of bacteria required for infection is also lower in those with reduced gastric acid, such as those using proton pump inhibitor drugs for ulcer or heartburn treatment, including commonly used drugs like Omeprazole.

TRANSMISSION AND SYMPTOMS

Cholera is usually transmitted through the faecal–oral route of contaminated water or food caused by poor sanitation. Symptoms normally begin suddenly, between twelve hours and five days after the bacteria is ingested. Normally, there is profuse watery diarrhoea and there may also be vomiting of clear fluid. Since an untreated person can produce between 10 and 20L of diarrhoea a day, this can result in a fall in blood pressure, a weak and rapid pulse and reduced urine output. Severe cases may also feature muscle cramps, altered consciousness, seizures or even coma, with the most dangerous complications being more common in children. The diarrhoea is often described as being like 'rice water' and may have a fishy smell, with a single diarrhoeal event causing up to a million-fold increase in the number of the vibrio cholerae bacteria in the environment.

One factor that has been found to influence infectivity is that upon excretion into the external environment, the bacteria themselves undergo a 24-hour period of hyper-infectious activity which helps to explain the explosive nature of cholera epidemics.

It is interesting to note that only humans are impacted by cholera and that the disease can only be passed between humans via infected water or food. In most developing countries the source is usually dirty water, while in wealthier countries transmission can occur when people harvest seafood, such as oysters, within water that is contaminated with sewage.

DIAGNOSIS AND TREATMENT

Where cholera is suspected, a rapid dipstick test is available, and where it is positive, stool and swab samples should be collected for laboratory analysis, especially before any antibiotics are prescribed.

The most immediate need is for patients to get their fluids rebalanced and one of the most dangerous errors in treatment is to underestimate the speed and volume of fluid that must be administered. Commercially produced solutions for rehydration are widely available but may be too expensive or unavailable, particularly when there is a sudden demand for treatment during an epidemic or in rural and undeveloped areas. A common recipe for an alternative to medicine is 1,000 ml of boiled water, half a teaspoon of salt and six teaspoons of sugar. More than one medical student has used such a formula for a bad hangover! Since potassium may be lost during episodes of severe diarrhoea, high potassium foods such as coconut water or bananas can be given, but carbonated drinks should be avoided as high levels of sugar can potentially interfere with water uptake. In severe cases, intravenous (IV) rehydration may be required.

Antibiotics can reduce the severity of the disease and reduce the length of its course, as well as decreasing the need for fluid replacement. Doxycycline is often used as a first-line treatment, as well as erythromycin, co-trimoxazole and chloramphenicol. Ciprofloxacin has also been used, but increasing resistance has emerged. This growing antibiotic resistance is a burgeoning hazard for our ability to effectively respond to sudden outbreaks of the disease.

PREVENTION

The best way to prevent cholera is through advanced water treatment and proper sanitation practices around the world, which will involve considerable investment in developing countries, especially in rural areas, a subject which is discussed later.

Several safe and effective oral vaccines are currently available. The WHO has three pre-qualified oral vaccines: Dukoral, Shanchol and Euvichol. To be effective, two doses are required and must be at least seven days apart.

For adequate protection when travelling to endemic and epidemic areas, the course must be finished at least one week before travel. If you happen to find yourself in an area where there is a sudden outbreak of cholera then extra preventive care needs to be undertaken: regularly washing your hands, only eating foods that are completely cooked (hot, not cold), only eating fruit and vegetables that you can peel yourself and ensuring that you only drink safe water, such as that which has been boiled or contains sterilising tablets.

CHAPTER 15

WATERBORNE DISEASES

The Pacific Institute described four classes of water-related diseases. From our perspective, the most important are waterborne diseases and water-based diseases.

Waterborne diseases result from ingesting water that is contaminated by human or animal faeces, or urine that contains pathological bacteria or viruses. This group includes some of the most catastrophic illnesses to hit humanity, including cholera, typhoid and dysentery (both amoebic and bacillary).

Water-based diseases are caused by parasites that are found in intermediate organisms, which themselves live in contaminated water. Examples from this group include schistosomiasis which affects over 200 million people worldwide and is responsible for over 200,000 deaths per year.

Another group, water-related diseases, include those illnesses that are spread by insects that breed or feed close to contaminated water. While these diseases, such as malaria, dengue fever and yellow fever are not directly associated with either the lack of access to clean water or sanitation, they can be associated with the construction of water systems, such as canals or dams, that create favourable conditions for their insect hosts.

A further consideration is the group of human illnesses caused by other forms of polluted water, mainly industrial pollutants, which is examined in the fourth section of this book.

THE CURSE OF TYPHOID

Typhoid is a waterborne disease caused by Salmonella typhi bacteria.

While cholera has featured in European medical literature since the 1600s, it is likely that typhoid has been around much longer. In 2006, remains from a mass burial site in Athens revealed fragments of DNA similar to contemporary Salmonella typhi DNA. The burial site contained the remains of those who died during the Plague of Athens which occurred in 430 BC, the second year of the Peloponnesian War between Athens and Sparta. The Athenians had been forced to withdraw inside the walled city to defend themselves but were soon short of drinking water, and the lack of basic sanitation led to widespread disease. Although many people believed that the catastrophic illness was caused by plague, no fragments of Yersinia pestis (plague) or typhus bacteria were discovered, making typhoid fever the hot favourite as the cause.

This disease has ravaged the world and claimed the rich and famous among its victims. In 1841, only thirty-two days into his term, the ninth President of the United States, William Henry Harrison, died of typhoid, making him the record holder for the shortest time in office. In 1861, Albert, Prince Consort and Queen Victoria's husband died three weeks after contracting the illness, leaving Victoria on the throne as a widow for another forty years.

The disease did not respect science either. Louis Pasteur, who saved countless lives by inventing vaccines for rabies and anthrax and who gave his name to the process of pasteurisation, lost his nine-year-old daughter Jeanne to the disease in 1859 and his twelve-year-old daughter Cecile in 1866. Another hero, Wilbur Wright,

died in 1912, aged forty-five, nine years after the world's first powered flight in the plane he co-invented with his brother Orville at
Kitty Hawk, North Carolina.

In 2015, over a century later, the World Health Organization estimated that there were between 11 and 21 million typhoid fever cases
worldwide, with around 150,000 deaths. A particular sub-type of
Salmonella typhi (known as haplotype 58) which is thought to have
originated in India in the late 1980s, now seems to be spreading
globally with a frightening level of multidrug resistance.

SIGNS AND SYMPTOMS

Unlike other strains of Salmonella, there are no known animal carriers of typhoid, with humans being the unique, and often asymptomatic, reservoir of infection. The typhoid bacteria grow in the
intestines, mesenteric lymph nodes, spleen, liver, gallbladder, bone
marrow and blood.

Infection, if untreated, has three main stages with each lasting
around a week, exactly the sort of disease pattern that history tells
us was exhibited by Prince Albert. In week one, although the patient's temperature gradually rises, it is not accompanied by the rise
in heart rate that is normally seen in fevers. Extreme tiredness, a
headache and cough are common symptoms and nosebleeds are
seen in around a quarter of cases. In week two, the patient's temperature will often reach a sustained level of 40°C (104°F) combined
with a continued slow heart rate and, often, the onset of delirium.
The spleen and liver become enlarged and tender to the touch,
rattling sounds are heard in the lower lungs and around a third of
patients exhibit 'rose spots' on the lower chest and abdomen. As the
disease worsens into the third week, the fever will remain high and
dehydration and malnutrition complicate the patient's delirium.

Intestinal haemorrhage can also occur, or worse, perforation, which may not reveal itself until symptoms of peritonitis or even widespread septicaemia develop. Pneumonia and encephalitis can also contribute to death. Up to 5 per cent of all those who contract typhoid fever and survive can become chronic carriers of the disease.

The good news is that when it is properly treated, typhoid fever is not usually fatal. Antibiotics, including amoxicillin, chloramphenicol and ciprofloxacin, have been used with widespread effect, although with the growing threat of antibiotic resistance, newer drugs like azithromycin or ceftriaxone may be needed. These will be combined with the general oral rehydration treatment given to those suffering from diarrhoea.

Diagnosis of the disease occurs either by culturing and identifying the bacteria from patient samples or by finding an immune response to the disease from blood samples. It is important to identify, by these means, those who may still be carriers of the disease, with gallbladder infection being a common focus in these patients, since carriers can knowingly, or unknowingly, be the source of further outbreaks.

THE TALE OF TYPHOID MARY

The notion of typhoid carriers was seared into the public consciousness due to the case of 'Typhoid Mary'. Mary Mallon was born in 1869 in County Tyrone, Northern Ireland, and since her mother had been infected with typhoid during pregnancy, it is thought she might have been born with the disease. After emigrating to the United States in 1884, she worked her way up to become a cook for prosperous New York families. From 1900, she worked for eight families in the New York City area, seven of whom contracted typhoid. In August 1906 she began a job in Oyster Bay on Long Island with another wealthy family who, again, developed typhoid. The public health authorities carried out their usual checks, but water samples taken from around the house, its pipes, toilets and cesspool all proved negative for the disease. It was then that a historic piece of medical detective work began, when private investigator George Soper was hired by the Oyster Bay property owner. He discovered that an Irish cook, who fitted a common description, had worked for all of the families involved. The cook was difficult to track down, as she left her employment each time an outbreak occurred. Eventually, Soper caught up with Mary, and after she was reported to the New York City Health Department, she was compulsorily quarantined. She was released in 1910, having given assurances that she would never work as a cook again and would improve her hygiene practices (she admitted to almost never washing her hands). Perhaps inevitably, however, the lure of a cook's wages was too much and she returned to her previous profession, resulting in further outbreaks. In 1915, she was returned to quarantine where she remained until her death in 1938.

Today, typhoid is a notifiable disease in the UK – it must be reported by law – and a reportable disease in the US where it must be reported within one working day of identification. There are currently around 400 cases of typhoid in the United States every year, around three-quarters of which are contracted during overseas travel.

In the UK, those who travel to areas of high risk, such as the Indian subcontinent, Africa, South and Southeast Asia or South America, are offered two main vaccines. Vi vaccine is administered as a single injection, while Ty21a is given as three capsules to take on alternate days. Although protection against typhoid lasts three years, as neither vaccine is 100 per cent effective, it makes sense to observe all the precautions on personal and food hygiene recommended for waterborne diseases in general.

It is worth ending this section by looking at one group of travellers who are not always volunteers – those in our armed forces. Typhoid vaccination has a particularly important place in their history, which is of special significance to me as both a doctor and former UK Secretary of Defence. In 1896, a British bacteriologist, Almroth Edward Wright, developed and introduced a typhoid vaccine which was used by the British Army in the second Boer War in South Africa. At that time, combatants were as likely to lose their lives to the disease as to battle, so the vaccine not only produced an improvement in personal health, but a significant boost to military capability. On the back of this success, at the outbreak of the First World War, Wright convinced the British Army to produce 10 million doses of the vaccine for the troops being sent to the Western Front. It is estimated that up to half a million lives were saved, with the British Army being the only fighting force to have its troops fully immunised at the outbreak of the war, taking on enemy troops, not typhoid.

POLIO

The story of polio is, in many ways, a tale of two worlds. In the developed world, the disease is almost eradicated. In parts of the developing world, especially Afghanistan and Pakistan, the disease remains endemic. How is this happening in the twenty-first century and how can we turn the situation around?

Polio, the common term for poliomyelitis, results from infection by poliovirus, one of a group of RNA viruses that colonise the gastrointestinal tract. In those with normal, functioning immune systems, infection is often asymptomatic, with around a quarter of cases only developing mild symptoms such as a sore throat or a mild fever. This situation will normally resolve within a week or two. Those who are infected may, however, continue to spread the disease for up to six weeks even if they have no symptoms. The disease is hugely infectious and spread through the faecal–oral route (by ingesting water or food contaminated by human faeces) or via the oral–oral route.

In around 1 per cent of infections the virus can migrate into the central nervous system from the gastrointestinal tract. Even then, most patients with central nervous system impairment will not progress to paralytic disease but will show symptoms of aseptic meningitis – headache, fever, neck and back pain, vomiting and lethargy.

Unfortunately, between 0.1 and 0.5 per cent of cases will go on to develop paralytic disease in which muscles, most commonly the legs, become increasingly weak and poorly controlled before becoming completely paralysed. Between 2 and 10 per cent of those who develop paralysis will die as the breathing muscles become affected. Three serotypes of poliovirus have been identified, with type II and type III certified as having been eradicated in 2015 and 2019 respectively. Type I (PV one) is not only the most commonly found variant but the one most associated with paralysis.

Polio only affects humans and, although there are treatments, there is no cure for the disease. Analgesics can be prescribed for pain, antibiotics to prevent infections in weakened muscles, and in severe cases, ventilation can be provided for those who cannot breathe independently. Extensive rehabilitation is often required. Patients who develop aseptic meningitis can expect a complete recovery, with symptoms lasting between two and ten days. In cases of spinal polio, some cells may temporarily lose function, but not be completely destroyed and they can recover within four to six weeks after the onset of the disease. Half of the patients with spinal polio will never recover, with patients equally split between those left with mild lifelong disability and others with much more serious lifelong damage.

A TALE OF THE DEVELOPED WORLD

In some ways, the history of polio infection seems to run counter to the narrative that we have followed, that increased sanitation leads to diminishing levels of disease. The great polio explosion actually followed widespread improvements in sanitation. For those living in the nineteenth century, polio was seldom seen in children under six months of age, with most cases occurring in children between six months and four years old. In those early days, sanitation was generally poor and meant that people were widely exposed to the virus with the consequent development of natural herd immunity across the population. Those under six months, who would normally be breastfed, would acquire protection from their mothers.

In the later part of the nineteenth and the early twentieth century, the world's more affluent countries began to deliver improvements in sanitation with the availability of clean water and proper sewage disposal. This had the effect of reducing exposure to poliovirus, and over time, the population lost its previously high levels of immunity. Around the turn of the twentieth century, small polio epidemics began to appear in the United States and Europe and by the 1950s the disease had reached pandemic proportions in North America, Europe, Australia and New Zealand. In the United States, in 1952, almost 58,000 cases were reported, with over 3,000 dead and over 21,000 left with mild to disabling paralysis. The peak age incidence of paralytic polio also departed from its previous pattern to children aged from five to nine years, where the risk of paralysis was greater. For parents in those years, polio was the greatest terror they faced.

Franklin D. Roosevelt, the thirty-second President of the United States from 1933 to 1945, first experienced symptoms of polio in 1921 at the age of thirty-nine. Throughout his life he remained paralysed from the waist down, depending on leg braces and a wheelchair for his mobility. Although he, and his team, made strenuous efforts to conceal the illness in public, many believe that his strength of character and determination, which were to serve America so well in the Second World War, were shaped by overcoming the adversity of his illness.

By 1977, there were over a quarter of a million polio survivors in the United States alone, many with varying degrees of lifelong disability because of polio paralysis. The WHO estimates that there are somewhere between 10 and 20 million polio survivors worldwide. While the disease gave rise to revolutions in intensive care and rehabilitation therapy, the positive consequences of which still benefit us today, the real watershed came with the development of the polio vaccine which could safely create the population immunity again.

The man responsible for this great milestone in public health is one of my personal heroes, Jonas Salk, an American virologist from New York City. It is not just Salk's great medical achievements that distinguish him but also his public spirit and noble personal attributes. In 1947, as polio rates in the population continued to rise, Salk became a professor at the University of Pittsburgh School of Medicine where, for the next seven years, he completely devoted himself to developing a polio vaccine. The results of his work were revealed to huge public and media acclaim in April 1955, and by 1959, it had reached around ninety countries. Salk's inactivated polio vaccine was based on poliovirus grown in monkey cell culture, which was then chemically inactivated with formalin. After two doses of the vaccine, 90 per cent of people will develop antibodies to all three types of poliovirus, with over 99 per cent developing immunity after three doses.

Another version of the vaccine, an attenuated live oral polio vaccine, was developed by Albert Sabin and came into commercial use in 1961. This type of vaccine uses a weakened version of the virus that causes disease, and because the vaccines are so like the natural infection, they stimulate a strong and long-lasting immune response. Today, these kinds of vaccines are used for mumps, measles, rubella and yellow fever, among many others.

Despite the enormous potential for profit, Jonas Salk chose not to patent the vaccine to maximise its availability and global distribution. When asked 'who owns this patent?', Salk replied, 'There is no patent. Could you patent the sun?' It is thought that if Salk had decided to patent his vaccine it would have been worth some $7 billion to him and his descendants. A man who genuinely disliked publicity, which he believed to be 'inappropriate for a scientist', Salk campaigned for the rest of his life on what he described as the 'moral commitment' of mandatory vaccines to prevent illness in children. He has inspired me by two of his most memorable quotes. The first is, 'I have had dreams and I have had nightmares. I overcame the nightmares because of my dreams', which I used to open *Rising Tides*. The second is, 'the reward for work well done is the opportunity to do more. Hope lies in dreams, in imagination and in the courage of those who dare to make dreams into reality.'

All too often we use the word 'inspirational' too loosely. Salk truly fits the definition.

A TALE OF THE DEVELOPING WORLD

In 1988, the Global Polio Eradication Initiative was launched by a coalition that included national governments, the World Health

Organization, Rotary International, the US Centers for Disease Control and Prevention (CDC) and UNICEF. The Bill and Melinda Gates Foundation and Gavi, the Vaccine Alliance, joined later. Their aim was the total global eradication of polio following the previously successful elimination of smallpox.

Enormous strides were made and, in 2000, polio was declared to have been eliminated in thirty-seven Western Pacific countries, including China and Australia. In 2014, the World Health Organization announced that it had now also been eradicated in the Southeast Asia region, including in Bangladesh, India, Indonesia, Sri Lanka and Thailand.

Wild poliovirus cases have decreased by over 99 per cent since, from an estimated 350,000 cases in more than 125 endemic countries to six reported cases in 2021, since the beginning of the project in 1988. While this is astonishing progress by any standards, it is not quite the end of the story. As of 2022, endemic wild poliovirus type 1 remains in two countries: Pakistan and Afghanistan.

The key problem is that until poliovirus transmission is interrupted in these countries, all countries remain at risk of the importation of polio, especially vulnerable countries with the dual issues of weak public health and immunisation services and those with travel or trade links to endemic countries. Continued internal disruption in Afghanistan with subsequent large volumes of migration, coupled with the impact of individuals harbouring extremist Islamic ideologies have made the journey towards eradicating polio more difficult.

Afghans constitute one of the largest global refugee populations. There are 2.6 million registered Afghan refugees in the world, of whom 2.2 million are registered in Iran and Pakistan alone. Another 3.5 million people are internally displaced, having fled their homes to search for places of safety within the country. As discussed earlier, Afghan migrants have increasingly spread to other parts of the globe, including the United Kingdom, arriving on small boats from Continental Europe.

A major complicating factor in the elimination of polio in Afghanistan and Pakistan, as well as in Nigeria, has been religious opposition, from a number of Islamic fundamentalists, to the delivery of immunisation programmes.

In 2003, in the northern Nigerian states of Kano, Zamfara and Kaduna, political and religious leaders called on parents not to allow their children to be immunised and effectively brought the immunisation campaign to a grinding halt. These leaders argued, without evidence, that the vaccine could be contaminated with anti-fertility agents, HIV and cancerous substances. As late as 2013, nine health workers administering the polio vaccine were killed by gunmen on motorcycles in Kano. The problem had arisen due to propaganda promoted by a physician in the city who was also the head of a prominent Muslim group, the Supreme Council for Sharia in Nigeria (SCSN).

He stated that the vaccines were 'corrupted and tainted by evildoers from America and their Western allies' and that 'we believe that modern-day Hitlers have deliberately adulterated the oral polio vaccines with anti-fertility drugs and … viruses which are known to cause HIV and AIDS'.

Echoing this conspiracy theory, the local Taliban in Pakistan and Afghanistan issued 'fatwas', denouncing vaccination as 'an American ploy to sterilise Muslim populations'. The Taliban subsequently assassinated vaccination officials, including Abdul Ghani Marwat, who was the head of the government's vaccination campaign in Bajaur Agency in the Pakistani tribal areas.

In February 2014, as many of the world's poorest children were unnecessarily being put at risk of polio, help to resolve this particularly dangerous situation arrived in the form of the world's leading Islamic scholars, led by the Grand Imam of the Holy Mosque of Mecca. They tackled the damaging propaganda by proclaiming that protection against diseases was not only admissible under Islamic Sharia, but obligatory and that any actions which interfered with preventive measures were, themselves, un-Islamic. The scholars adopted a strong 'Jeddah Declaration' and produced a six-month plan of action to address critical challenges facing the campaign to eradicate polio in the polio-endemic parts of the Islamic world.

The episode is a clear reminder of the danger that ill-intentioned or ill-informed individuals or organisations can pose to immunisation programmes. The rise in cases of measles in western countries, resulting from a diminished uptake of the MMR vaccination can be attributed to damaging and disproved claims about links to autism. The rise of the anti-VAX conspiracy theorists after Covid-19 shows that the risk is ever-present and is now magnified by social media.

2022

While the oral polio vaccine is highly effective, it possesses one major disadvantage in that it contains a live virus. Although this has been attenuated, it can still spread from person to person and retains the ability to mutate into something more serious.

The danger posed by the failure to eliminate polio in countries where it remains endemic was highlighted in 2022, and it underlined the point that until such diseases are eliminated everywhere, they remain a potential risk to all.

The oral vaccine is the mainstay of the global eradication programme because it is cheap, easy to use and because the level of immunity conferred is good. One drawback is that when immunisation rates are low, the vaccine virus can spread from person to person and eventually change enough, through mutations, to regain its ability to paralyse. This is known as vaccine-derived poliovirus (VDPV). Alarm bells sounded, almost certainly magnified by media hype, in London in 2022, when the virus was detected at two points in London's sewage system and was later confirmed to be vaccine-derived poliovirus. Similar findings occurred in New York with one case of paralysis reported in New York State. A genetically similar virus has also been found in Jerusalem. Professor Nick Grassly, of Imperial College London, explained that the attenuated virus, because it grows in the intestine for a short period, can be detected in faeces. The most likely explanation for the contaminated water in London was that, he said, 'in recent months there have been campaigns to attempt to eradicate polio from Afghanistan and Pakistan, and it could well be a traveller from one of those countries inadvertently bringing the virus in the gut and then shedding that virus into our sewers'. Experts agree that large outbreaks of polio remain very unlikely in wealthy countries due to good sanitation and high vaccine coverage rates. Yet, there is no room for complacency. As Professor Grassly pointed out,

> We need to ensure that children are up to date with their vaccines. In London, 13 per cent of infants aged twelve months may not have been fully vaccinated against polio, and in some London boroughs, this number is substantially higher, reaching 32 per cent in Hackney and the City of London. Failure to be vaccinated puts these children at risk of polio paralysis.

The lesson is clear – if we want to protect our populations and improve our general immunity to disease, we must begin by protecting our children and that means vaccinating them against diseases such as polio, so that we never again have to endure the scourge and fear that our predecessors did.

CHAPTER 16

THE NEGLECTED
TROPICAL DISEASES

Accoring to the WHO, neglected tropical diseases (NTDs), are a diverse group of twenty conditions, mainly affecting tropical areas and which have a disproportionate effect on women and children. The diseases are 'neglected' because they affect the poorest in the developing world, because funding priority has been given to HIV/AIDS, malaria and TB, and because the social stigma around these diseases is likely to prevent individuals seeking help and sticking to treatment. Since this is not a medical textbook, we will only look at a handful of the most significant and destructive diseases and keep their descriptions to a minimum, but I hope this gives an idea of the scale and variety of the human cost incurred.

In January 2012, some of the world's leading pharmaceutical companies and NGOs signed the London Declaration on Neglected Tropical Diseases which focused on controlling or eradicating ten of these diseases by 2020. The aim was to produce a package of drugs that would simultaneously target some of the most common human helminthiases – worm infections due to tapeworms, flatworms (flukes) and roundworms. These infect more than a quarter of all the people on our planet at any one time, a far greater number than HIV and malaria put together. Schistosomiasis, for example, is the second most prevalent parasitic disease in the world after malaria, with around 200,000 people dying every year from conditions related to the disease. These conditions, which could be destroyed by good sanitation and access to clean water, are a stain on our record as truly civilised human beings in the twenty-first century.

THE HELMINTHS

Helminths are parasitic worms which mostly live in their hosts' gastrointestinal tract but can cause damage to a range of other organs if they burrow into them.

During our (by historical standards) short evolutionary history and our journey out of Africa, humans have acquired an amazing number of parasites, including about 300 species of helminth worms.

A set of ancient Egyptian documents, known as the Ebers Papyrus, refer to intestinal worms, and calcified helminth eggs have been found in mummies dating from around 1,200 BC. The Greek physician Hippocrates, who is often regarded as the father of medicine, knew about worms from fishes, domesticated animals and humans and while Roman physicians also seem to have known about human roundworms and tapeworm infections, the science of helminthology only took off in the seventeenth and eighteenth centuries following the re-emergence of science and scholarship during the Renaissance period.

One of today's experts, Professor F. E. G. Cox from the Department of Infectious and Tropical Diseases at the London School of Hygiene and Tropical Medicine divided them according to their role in our evolutionary journey as either 'heirlooms' or 'souvenirs'. Heirlooms are the parasites inherited from our primate ancestors in Africa, and souvenirs are those that we have acquired, along our subsequent journey, from the animals with which we have come into contact during our evolution, migrations and agricultural practices. The wider transmission of infections between humans was facilitated by the development of settlements and cities, and the opening of trade routes resulted in an even wider dissemination of parasitic infections. Among its many other evils, the slave trade, which flourished for three and a half centuries from approximately 1500, brought new parasites from the Old World to the New World where another, less commonly discussed, legacy remains.

TYPES OF INFECTIONS

We can divide these types of infections into two main groups: the roundworms (nematodes) which include the helminths transmitted by soil and the filarial worms, along with the flatworms which include flukes (trematodes) and tapeworms.

HELMINTHS

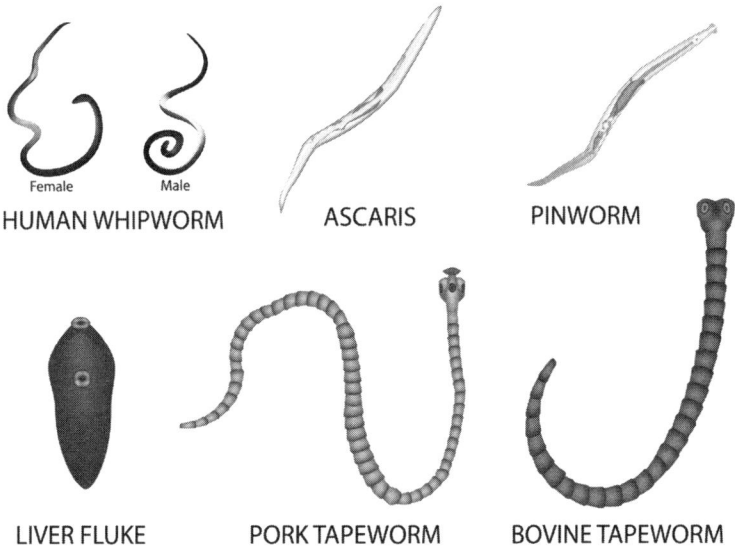

Female Male
HUMAN WHIPWORM   ASCARIS   PINWORM

LIVER FLUKE   PORK TAPEWORM   BOVINE TAPEWORM

Children tend to harbour the highest numbers of intestinal worms and schistosomes and may therefore be subject to gross stunting, impaired memory and diminished cognitive function, locking the legacy of these diseases into future generations. Hookworms and schistosomiasis result in neonatal prematurity, diminished birth-weight and increased maternal morbidity and mortality. Onchocerciasis is a leading cause of blindness and skin disease, while lymphatic filariasis (LF) is a major cause of limb and genital deformities. The economic consequences of this level of chronic disease mean that many of the countries where these diseases are prevalent are inevitably stuck in endless cycles of poverty, chronic illness and death.

So, if the problem is of such magnitude, why do we have so few weapons at our disposal? The answer probably lies in a combination of our lack of detailed knowledge about helminthic infection and its interaction with human disease and genetics on one hand, and the very modest commercial opportunities that currently exist for drugs targeting human helminth infections. In fact, the study of these diseases receives less than 1 per cent of global research finance.

These kinds of infections differ markedly from some of the disease entities we have previously considered, including viruses and bacteria, because helminths do not replicate within the human host. The severity of the disease is therefore linked to the number of worms infecting the host, i.e., the worm burden, rather than the absence or presence of infection.

Helminths are most frequently transmitted to human hosts through the ingestion of contaminated vegetables, drinking water and raw or undercooked meat, with soil-transmitted helminths highly affected by surface temperature, altitude, soil type and rainfall.

Diagnosis is possible through microscopic examination of the eggs found in human faecal samples. The general pattern of the life cycle can clearly be seen by examining the case of the ascariasis roundworms which cause the disease ascariasis.

ASCARIASIS

Most readers may have never heard of this disease, but ascariasis is one of the most common worm infections in the world, with estimates suggesting that between 1 billion and 1.5 billion people are affected, in other words, around one in seven people on the planet, the equivalent of around three times the population of the entire European Union.

Ascariasis worms are typically pink or white with tapered ends. Female worms can be more than 15 inches (40 centimetres) long and a little less than a quarter of an inch (6 millimetres) in diameter. Male worms are generally smaller. Normally, adult worms will live in the intestines until they die and most people who are infected will have mild cases with no symptoms, or mild symptoms ranging from vague abdominal pain to diarrhoea or bloody stools. Heavy infestation, however, can lead to much more severe illness. When the eggs hatch in the small intestine, the larvae can migrate through the bloodstream or the lymphatic system into the lungs, producing a pattern of illness like pneumonia or asthma, with breathlessness, wheezing and a persistent cough. A heavy intestinal load may lead to severe pain, weight loss or malnutrition and a worm may also appear in the patient's faeces or even their vomit.

SCHISTOSOMIASIS

Schistosomiasis can present as both an acute and chronic disease and is caused by parasitic flatworms called schistosomes. It is also known as snail fever, Katayama fever (from the Katayama district of

Hiroshima in Japan, where schistosomiasis was once endemic) and bilharzia (after the German pathologist Theodor Bilharz, who first discovered the worms while working in Egypt in 1851).

In 2019 the disease affected about 240 million people worldwide and estimates of deaths range from 4,500 to 200,000. It is most commonly found in Africa, Asia and South America, with around 700 million people at risk in more than seventy countries where the disease is common.

People are infected when the larval forms of the parasite, which are released by specific freshwater snails, penetrate the skin when it is in contact with infested water. People with schistosomiasis then contaminate freshwater sources with urine or faeces containing parasite eggs which then hatch in the water and continue the cycle of transmission. Inside the body, the larvae develop into adult schistosomes. These live in the blood vessels where the females release eggs, some of which pass out of the body again to continue the parasite's life cycle while others can become trapped in the body's tissues, causing progressive damage to organs and immune reactions.

The disease is prevalent in tropical and sub-tropical areas, especially those without access to safe drinking water or proper sanitation. Around 90 per cent of people requiring treatment for schistosomiasis live in Africa, mainly in poor and rural communities. Agricultural and fishing populations, being frequently exposed to infected water, are particularly vulnerable, as are women performing domestic chores, such as washing clothes in infested waters, who can also develop female genital schistosomiasis.

As rural populations migrate to urban areas the disease is introduced to new populations, the same phenomenon that occurs when tourists, keen to have new travel experiences, explore more remote areas with increasing numbers of them contracting schistosomiasis.

As with other waterborne diseases, methods of preventing the spread includes improving access to clean water but, in this case, a key element is reducing the number of snails. The drug Praziquantel may be given to entire groups of the population who are affected once a year.

Many affected individuals will only suffer mild symptoms, including a light or tingly rash, often known as swimmer's itch, about twelve hours after infection.

More severe symptoms can occur between two and ten weeks later and include fever, cough, muscle aches, diarrhoea and enlargement of the liver and spleen. The disease usually self-resolves in two to eight weeks with steroids used to reduce inflammation and Praziquantel to kill the adult schistosomes.

In chronic disease, inflammatory reactions can cause long-term damage to the intestines, liver, spleen and urinary systems and is also associated with an increased risk of bladder cancer. Theodor Bilharz would have no doubt been gratified to know that successful efforts to control schistosomiasis in Egypt have led to large decreases in the rate of bladder cancer there in recent years.

The life cycle of the schistosome has an interesting link to the sort of dam and watercourse engineering that we discussed earlier. Infected individuals, as we have seen, release eggs into the water via their faeces or urine and, after the larvae hatch from these eggs, they infect freshwater snails. After reproducing and developing, the parasites leave the snails and enter the water system where they can only survive for around forty-eight hours without the mammalian host.

Ever since the 1950s, the United Nations has set out planning and building specifications designed to avoid dams and irrigation schemes producing a rise in waterborne schistosomiasis infections.

Irrigation projects can be designed to make it difficult for the snails to colonise the water and to reduce the contact with the local population. Unfortunately, many of these guidelines were either unknown or have been ignored, although there have been a few notable successes where rebalancing the natural habitat has had spectacular results. Since the dams have reduced the population of a large migratory prawn which eats the snails, simply restoring the prawn population upstream can reduce both the number of snails and the human reinfection rate as has successfully occurred in the Diama Dam on the Senegal River, which spans the border between Mauritania and Senegal.

The main line of defence against schistosomiasis, along with improvements to water systems, is to take a single of Praziquantel annually, although this only eliminates adult schistosomes and is not effective in killing the eggs and immature worms. When a village reports more than 50 per cent of the children in an affected area as having blood in their urine, everyone in the village now receives treatment under WHO guidelines.

In geopolitics, the parasites have played a curious historical role in the complex relationship between China and Taiwan. In 1949, as the Japanese Empire was being dismantled after the Second World War, the island of Taiwan, then known as Formosa, was ceded to the Chinese authorities (having previously been ceded to the Empire of Japan by China in 1895). The nationalists under General Chiang Kai-shek occupied the island, and the communists under Mao Zedong began preparations for the invasion of the island by the People's Liberation Army (PLA). An important part of the preparations for the amphibious assault was a programme of swimming lessons for the Chinese troops, the majority of whom had never lived near water. Unfortunately for them, the streams and canals chosen for the exercise, throughout the southeast provinces, were densely inhabited by snails along with their parasites, *Schistosoma japonicum*.

A few weeks after beginning their swimming exercises, PLA soldiers began to fall ill with skin rashes, fever and severe cramping. With the onset of what became the world's largest acute epidemic of schistosomiasis, which is estimated to have affected between 30,000 and 50,000 soldiers, there would be no invasion of Taiwan. The schistosomes played a key role in ensuring its independence which, of course, remains threatened by China to this day.

LYMPHATIC FILARIASIS

This disease is mentioned, though it is not technically a waterborne disease, due to its symptomatic association with water. Lymphatic Filariasis (LF) was the first mosquito-borne disease to be identified and is caused by filarial roundworms which attack the lymphatic system. Around 40 million people are currently infected with the disease, and around 880 million people are at risk in fifty-two countries worldwide. The worms lodge in the lymphatic system, obstructing lymphatic flow and causing different tissues, depending on the specific type of worms, to swell. Arms, legs, breasts and genitals can all be affected. The disease can lead to the condition known as elephantiasis, where the disrupted lymphatic flow and swelling of the skin and underlying tissues can create an elephantine appearance.

The WHO is co-ordinating a worldwide eradication of the disease using insect repellents and mosquito nets infused with insecticides. This is combined with the mass deworming of populations who are at risk with the drugs Ivermectin and Albendazole. Sri Lanka was certified as LF-free in 2016, joining a club that includes Cambodia,

China, Egypt, the Maldives, South Korea, Thailand and Vietnam. The target for 2030 is to eliminate LF in 80 per cent of endemic countries which would represent a huge reduction in the miserable burden currently endured by so many millions.

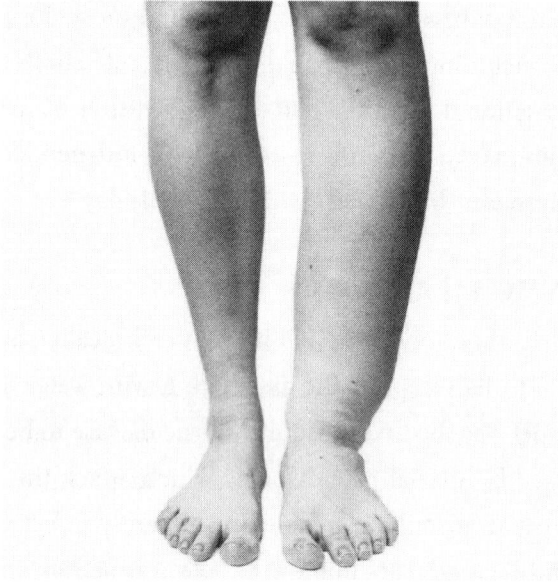

TRACHOMA

Trachoma, caused by the bacterium Chlamydia trachomatis, causes more loss of vision and complete blindness than any other infection in the world. Almost 1.9 million people have lost their vision because of the disease. It causes 1.4 per cent of all blindness worldwide with 136 million people living in trachoma endemic areas at the highest risk. Children are affected more than adults as they touch their face more frequently and women are four times more likely to suffer from trachoma than men, largely due to their traditional roles in caring for children, who are often the primary host for the disease. It is spread from the fingers, via clothes and towels and from flies moving from person to person. It is particularly prevalent, as with other waterborne diseases, where there is a lack of access to clean water and proper sanitation. If untreated, the eyelashes will scratch the cornea, causing severe pain and potentially irreversible blindness. Where blindness has not completely progressed, surgery can reduce the pain and stop the progression of the loss of vision. Antibiotics, including Zithromax (azithromycin), which is donated to the International Trachoma Initiative by Pfizer, can, alongside facial cleanliness, washing the hands with soap and the face with water, result in major improvements.

DYSENTERY

Dysentery is a type of gastroenteritis that produces bloody diarrhoea and is usually caused by either Shigella bacteria (shigellosis) or the amoeba, Entamoeba histolytica (amoebiasis or amoebic dysentery). Shigella is estimated to cause around 1.1 million deaths a year with over 50,000 of those among children. Amoebiasis infects over 50 million people a year, of whom one in one thousand will die.

Bacillary dysentery is the more common of the two forms and usually results in a mild sickness which presents one to three days after infection. While symptoms normally consist of slight abdominal pain and diarrhoea, extreme cases can result in the loss of more than a litre of fluid per hour, the effects of which can be understood due to the earlier section on human water balance. Treatment, as with many similar diseases, is by managing fluids via oral rehydration, although IV fluids may be required in severe cases. In these, patients will present with diarrhoea containing blood, have extreme abdominal and rectal pain and a low-level fever. Anaemia can also occur due to blood loss through diarrhoea. Antibiotics can be used but many strains of Shigella are becoming resistant to the more common drugs, another example of where resistance may prove a major problem in fighting disease in the future.

Amoebic dysentery is mostly found in tropical areas, especially where human faeces is used as fertiliser. Proper treatment, with effective antibiotics is important, as untreated disease can lead to severe complications years later.

Throughout the centuries, dysentery has been no respecter of rank. In British history, King John (of Robin Hood fame), Edward I (Hammer of the Scots) and Henry V were all high-profile royal victims of the disease, an indication of how long it has been taking its grim human toll.

E. COLI

E. coli bacteria are normal residents in human intestines and usually colonise an infant's gastrointestinal tract within two days of birth, arriving with water or food or from those handling the child. Non-pathogenic E. coli may even be beneficial to human health by producing vitamin K2 or preventing other pathogenic bacteria from seizing control of the gut.

Most E. coli strains do not result in disease, but a few pathogenic sub-types can produce serious illness.

Enterotoxigenic E. coli (ETEC) is the most common cause of diarrhoea in travellers, with around 850 million cases arising in developing countries each year. Some strains, for example O157: H7 can produce the Shiga toxin, resulting in gastroenteritis with bloody diarrhoea, urinary tract infections, neonatal meningitis and Crohn's disease. Treatment is with fluid rehydration, although antibiotics may be required in more severe cases of infection.

In the UK in 2023, an outbreak of the Shiga toxin, producing E. coli, O183 serotype, produced twenty-five severe cases of illness with one death. Although this is a rare serotype, with only fifteen cases having been identified in the UK since 2016, public health authorities are quick to react to find potential sources and limit the spread. Its detection is a priority for public health authorities, not just in developed countries, but everywhere.

DRACUNCULIASIS (GUINEA-WORM DISEASE)
Not everything is bad news, however.

Amid the gloom about the continued human toll of neglected tropical diseases, let's finish with a success story which will, hopefully, become the template for making major inroads against many of these terrible conditions.

Dracunculiasis is a parasitic infection caused by the Guinea worm (dracunculus medinensis), which is seldom fatal but can cause those who are infected to become non-functional for weeks and months with devastating social and economic impacts. It affects those living in rural, deprived and isolated communities who mainly depend on open stagnant surface water, such as ponds, for their drinking water. When this water, containing fleas infected with guinea worm larvae, is drunk, these larvae migrate from the digestive system to an exit site (usually a lower limb) where a painful blister appears. This eventually bursts and the worm slowly crawls out over several weeks. This phenomenon, echoed in numerous urban legends and horror stories, is no fantastical creation, but a reality, causing misery for millions.

Dracunculiasis in humans has been known about since 1,000 BCE at least, and by the nineteenth and early twentieth centuries, it affected as many as 48 million people per year across much of Africa and South Asia.

In the early 1980s, international agencies decided that they could, and should, move to eliminate the disease. During the mid-1980s an estimated 3.5 million cases of dracunculiasis occurred in twenty countries, seventeen of which were in Africa and three others in Asia. In 2007, the number of reported cases fell to below 10,000 for the first time and by 2012 it had fallen to only 542 cases. It has remained in double digits over the past decade, with cases in humans reported only in Angola, Chad, Ethiopia, Mali, South Sudan and Cameroon.

This crippling parasitic infection is now on the verge of eradication. It is possible to do it. Our task must be to provide the access to clean water, sanitation and basic healthcare that will consign other diseases to history. How to do it is examined in the next chapter.

CHAPTER 17

SANITATION AND THE RIGHT TO CLEAN WATER

One of the main reasons behind writing this book is my deeply held belief that every human being should have a right to clean water. Much of my time as a politician has been spent talking about other human rights which, while entirely justifiable and desirable in themselves, do not literally bestow the right to life itself.

Any society can only achieve high rates of public health, economic productivity, gender equity, educational attainment and a host of other desirable outcomes when all of its members enjoy their rights to water and sanitation.

In fact, the right to water is an element of 'the right of everyone to an adequate standard of living for himself and his family' (Article 11 of the International Covenant on Economic, Social and Cultural Rights [ICESCR]).

The right to water means that everyone is entitled to have access to sufficient, safe, acceptable, physically accessible and affordable water for their personal and domestic use.

The right to sanitation means that everyone is entitled to have physical and affordable access to sanitation that is safe, hygienic, secure, socially and culturally acceptable and that provides privacy and ensures dignity.

When we talk about such rights, it is important to stress that there is a difference between access and physical presence. A water or sanitation service does not serve the whole community if it is too expensive, unreliable, unhygienic, in an unsafe location, unmodified for less-able groups or children or is non-gender segregated, in the case of toilets and washing facilities.

The journey to this concept of a right to water has been a long one. In considering where many of our modern concepts come from, the experience of Victorian Britain is an interesting one to consider, since, as the country was at the height of the Industrial Revolution and the British Empire, its implications had widespread global repercussions.

VICTORIAN LONDON'S GREAT STINK

During the first half of the nineteenth century, London's population was ballooning and the city was the capital of the biggest empire the world had ever known. The problem was, from a public health perspective, that more people meant more waste. Authorities had tried, largely in vain, to keep abreast of the challenge. In the century before 1856, over a hundred sewers had been built and by this time the city contained around 360 sewers and 200,000 cesspits, some of which leaked methane which tended to catch fire and explode. Since most homes did not have flush toilets, 'night soil men' collected some solid waste for fertiliser but a great deal simply ended up lying on the city streets or in its watercourses. For those fortunate enough to have flush toilets, they simply displaced any sewage into the biggest sewer of all, the River Thames. The ever-increasing level of activity from London's booming economy also meant that an ever-greater amount of additional waste from factories and slaughterhouses, along with other industrial byproducts heaped added pressure upon a failing system. By 1857, the smell from the Thames was so bad that the government of the day ordered chalk lime, chloride of lime and carbolic acid to be poured into the river to alleviate the rancid odour. It was not, of course, a new problem. Over 300 years earlier, way back in 1535, an Act was passed to stop the dumping of excrement into the river, and long before the Industrial Revolution, it was described as thick and black due to sewage.

The scorching summer of 1858 brought things to crisis point. In June, with temperatures averaging 34 to 36°C (93 to 97°F) in the shade and up to 48°C in the sun, the dry conditions saw the river levels drop to such a low level that raw effluent from the city sewers lay on the riverbanks. It was the beginning of what became known as 'the Great Stink'. Charles Dickens wrote, 'I can certify that the offensive smells, even in that short whiff, have been of a most head- and stomach-distending nature'.

Britain was reaping the whirlwind of urbanisation that came in the wake of the Industrial Revolution and it was not just the stink but the cholera outbreaks of the nineteenth century, described earlier, that were the results of what was effectively an open sewer flowing through the capital city.

Sitting on the edge of the river, the recently rebuilt Houses of Parliament received the full force of the river's foulness, so much so that discussions about moving the government itself to Oxford or St Albans took place. The Prime Minister, Benjamin Disraeli, was reported to have rapidly left a committee room 'with a mass of papers in one hand, and with his pocket handkerchief applied to his nose' because of the putrid air. On 15 June he introduced new legislation to the House of Commons, the Metropolis Local Man-agement Amendment Bill, which described the River Thames as 'a Stygian pool, reeking with ineffable and intolerable horrors'. This assigned responsibility for cleaning up the river to the Metropolitan Board of Works which subsequently took control of the sewers. The legislation allowed the Board to borrow £3 million for the task, which was to be repaid from a three-penny levy on every London household over the following forty years. The plan adopted to put this project into effect was that of the civil engineer Joseph Bazalgette, who planned to build a series of interconnecting sewers that would take effluent eastwards and out into the Thames Estuary to be dispatched on the outgoing tide. The project started at the beginning of 1859 and was completed in 1875 at a cost of some £2.5 million (about £300 million in today's money). It replaced around 150 miles of ageing sewers with over 1,000 miles of new ones, creating, in the process, flood barriers in the form of the river embankments which are such a familiar feature of today's London for residents and tourists alike. Of course, not everyone was happy and those who lived downstream now had to contend with a new system where London's effluent entered the river much closer to where they lived and simply, in their view, displaced the problem from central Londoners to them. The compromise was a fleet of vessels which were commissioned to deposit the waste out at sea, a method of waste management which remained legal until it was banned in 1998.

Despite all the work by doctors, scientists and public health reformers of the time, it is arguable that Bazalgette did more to save lives than any of his contemporaries, dramatically improving sanitary conditions in the capital and striking a huge blow against its waterborne diseases.

Although there is a great deal of focus on events in London, meaningful legal reform in the area of sanitation had begun across the country in the preceding decades.

The 1848 Public Health Act was the first law on public health to be passed in the United Kingdom. Its main architect was the social reformer Edwin Chadwick who had been behind the passage of the 1834 Poor Law. His arguments were largely economic, that if the health of the poor were improved it would cost less in government support. The aims he sought to achieve through his public health reforms were improved drainage and the provision of sewers, the removal of all refuse from houses, streets and roads, the provision of clean drinking water and the appointment of a medical officer for each town. While the Act was farsighted, some contemporaries might have said revolutionary, and it established a central board of health, this organisation had limited powers and no money. In effect, although it provided a framework for local authorities, there was no compulsion upon them to take action.

In 1874, Benjamin Disraeli led the Conservative Party to power. It was committed to the improvement of public health as part of a wider programme of social change and it was decided that the 1875 Public Health Act would consolidate all the previous legislation relating to public health that had been passed during the nineteenth century.

The law established and named local authorities as rural and urban sanitary authorities which would replace the local boards of health that had been established in 1848. The local authorities were required to provide clean water, ensure that only safe food was sold and dispose of all sewage and refuse. As well as giving them the power to ensure that homes were connected to the main sewerage system, it outlawed the building of new homes that did not have such a connection.

Although the Act was seen as largely a consolidating measure, there was a clear desire on the part of many to regard it as a stepping stone to further measures. As the Right Honourable Lyon Playfair,

MP for the universities of Edinburgh and St Andrews noted at the second reading of the bill: 'Our public health in England is so low that we suffer annually 125,000 preventive deaths and have 3 million or 4 millions of serious cases of preventable sickness, weakening the industrial powers of the survivors.' Playfair would later go on to establish the Royal College of Science and South Kensington Museum in London. The Act represented the beginning of a political and public awakening about the importance of public health which would prove to be irreversible.

It is this link between the historical mission to achieve clean water and the levels of water purity and sanitation that we take for granted in the developed world today.

THE GLASGOW EXPERIENCE

Glasgow, the city in which I studied medicine, is a prime example of how access to clean water can transform both health and economic prospects. Having had a population of 32,000 in 1750, the city grew to over 200,000 a century later. As an important port with access to the Atlantic Ocean, Glasgow prospered by importing American tobacco and cotton along with Caribbean sugar, a trade which continued to flourish after the Act of Union with England in 1707, so that by 1760, Glasgow had outstripped London as the main British port for tobacco. This expansion was assisted by the desilting of the River Clyde in the 1770s, which not only allowed bigger ships to move further up the river but laid the foundations for the future shipbuilding industry in which 'Clydebuilt' would become a global benchmark for quality. By the 1840s Glasgow was a burgeoning city with the abundance of coal and iron in Lanarkshire enabling industrial development alongside the textile mills based on imported cotton and the availability of wool.

While all this economic development was positive in its own right, the city dramatically outgrew its water supply, most of which still came from an 1807 scheme which used the River Clyde as a source. As the quality of the water from the river declined and outbreaks of cholera became more frequent (an epidemic in 1848–49 killed over 4,000 people), the city council turned to water engineer John Frederick Bateman, who had already designed Manchester's water supply system, to provide them with a solution. Bateman's plan was to bring an unlimited supply of fresh water to the city from Loch Katrine in a project involving reservoirs, almost 26 miles of aqueduct, 13 miles of hard rock tunnels and almost 4 miles of iron pipes. Construction began in 1855 and the aqueduct was opened by Queen Victoria in 1859.

Loch Katrine is a freshwater loch in the Trossachs, east of Loch Lomond, and is around 8 miles (13 km) long and 1 mile (1.6 km) wide at its widest point. It is the fictional setting of Sir Walter Scott's poem 'The Lady of the Lake' and the subsequent opera, *La Donna del Lago*, by Gioachino Rossini. An astonishing work of civil engineering, the aqueduct was laid at a gradient of just 10 inches per mile, dropping just 1 m for every 6,334 m horizontally. From the north end at Loch Katrine to the south end at the town of Milngavie it appears almost flat. The true engineering genius can be seen in the fact that if pumping had to be done using electricity rather than gravity, it would require about 64 million kWh (kilowatt hours) per year. Bateman's great scheme transformed the cholera-afflicted city into one of the richest cities in the world by the late nineteenth century, with a municipal public transport system, internationally renowned museums, libraries, great parks and flourishing industry that would lead it to become known as the 'second city of the empire'.

THE EARLY CHALLENGES

From the beginning of civilisation, one of humanity's main challenges has been how to access clean water to drink and finding a suitable method of disposing of our waste. The use of surface water from rivers and the digging of wells were the main sources of water, as discussed in Part 1.

The oldest reliably dated well is from the Pre-pottery Neolithic (PPN) site of Kissonerga-Mylouthkia in Cyprus where a circular shaft was driven through limestone to reach an aquifer at a depth of 8 metres (26 ft). This has been dated at around 8600 BC.

The first stone-lined well, at a depth of 5.5 metres (18 ft), has been documented from a drowned PPN site off the coast near modern Haifa, in Israel, and dated at around 7000 BC, while in Europe, an oak water well discovered in the Czech Republic is the world's oldest known wooden construction, dating from approximately 5250 BC.

In ancient Egypt, the temple complex at Abusir and the associated pyramid, dated at around 2400 BC, was found to have a network of copper drainage pipes. The Minoan civilisation on the island of Crete not only had a system for bringing clean water into its capital Knossos, but also a system to take out wastewater, along with a complex system of canals to deal with storm sewage overflow. It also appears to have produced one of the first flush toilets in the eighteenth century BC. The Romans, of course, had indoor plumbing, both private use and public baths, which was made possible by their incredible engineering skills, particularly the use of aqueducts to transport freshwater over huge distances. The same mixture of public and private water supplies was seen in the Abbasid Caliphate, which existed from 750 to 1517 AD and where the capital, Baghdad, not only had 65,000 baths and a functioning sewer system but where hydraulic technology was used to ensure that even multi-storey buildings had both water supplies and sewage outlets. Much of this would have come as a disappointment to Sir John Harrington who, in sixteenth-century England, invented a flush toilet for Queen Elizabeth I, believing it to be a world first. Despite inheriting the water, plumbing and sewerage legacy of the ancient world, little was done to take the skills of the Roman Empire forward until the time of the Enlightenment. Advancements were then accelerated by the Industrial Revolution which saw large populations move to the cities and caused both an increased demand for water and an accompanying need to deal with sewage to prevent outbreaks of disease. Even before London upgraded its system, engineers in the English city of Liverpool had produced a groundbreaking system of sewers and drains. Ten years before the Great Stink in London, the Scottish engineer James Newlands had built almost 300 miles of infrastructure which contributed to doubling the life expectancy in the city.

In the late 1850s, new sewer systems came to Chicago and Brooklyn in the United States where an important innovation was created – water treatment using chemical precipitation – which first appeared in 1890 in Worcester, Massachusetts.

The early gravity-based sewers simply discharged their contents into surface waters, but around the turn of the twentieth century authorities began to treat the sewage itself to diminish pollution and the incidence of waterborne diseases. Their success led to increased life expectancy and, with it, a burgeoning population which, of course, added still more to the scale of the problem. The use of cesspools, in which solid waste slowly liquefied due to anaerobic digestion, gave way to septic tanks where bacteria were used to accelerate the process. In the 1870s, on his sewage farm in Croydon, England, Edward Franklin showed that filtration of sewage through porous gravel created a nitrifying effluent and that the filter remained un-clogged for long periods. This led to the more generalised treatment of sewage by biologically decomposing the organic components using bacteria. A greater concentration of biological organisms was created by the British engineers E. Ardern and W. T. Lockett in 1913 in what they called the activated sludge process and the first full-scale continuous flow system was installed in Worcester, England in 1916. After the First World War, new technology spread quickly through the USA, Canada, Germany, Denmark and other European countries as attention turned to using science to improve living standards for the post-war population.

Treating the water itself to improve purity had long been an aim of public provision. Experiments with filtering water through sand culminated in 1829 when a fully filtered public water supply was provided by the Chelsea Water Works Company in London, a system that was replicated throughout the United Kingdom in the following decades and accelerated by the demands of increasingly vigorous legislation. The permanent chlorination of water began in 1905 when Dr Alexander Cruikshank Houston chlorinated water to prevent a serious typhoid epidemic in Lincoln, England, while the first continuous use of chlorine in the United States occurred for the water supply in Jersey City, New Jersey, from the Boonton reservoir in 1908. These developments continued, coming together and being constantly updated with new technologies until they became the water and sanitation services that most of us in the developed world take for granted today.

WASH AND DEVELOPMENT GOAL 6
While in the developed world we have come to take our clean water supplies, efficient sanitation and sewage treatment systems for

granted, the same cannot be said for many of our global neighbours in the developing world and the international focus has increasingly been on how to close that gap.

One of the most common acronyms in the current debate on global improvements in this field is WASH, which stands for water, sanitation and hygiene. The first two objectives for the UN's Sustainable Development Goal 6 (SDG 6) are equitable and accessible water and sanitation for all. UNICEF estimates that, in 2022, 3.5 billion people still lack safely managed sanitation services and 844 million people were living without access to safe and clean drinking water. The difference between the most developed nations with the kind of advanced water and sanitation described above and the millions who have fallen behind can be seen in the mortality rates associated with the gap. The connection between the lack of WASH and the burden of disease is a clear reflection of the poverty and poor access in developing countries. The World Health Organization has estimated that the WASH attributable mortality rates per 100,000 population were 3.7 in high-income countries, 4.4 in upper-middle-income countries, 30 in lower-middle-income countries and 42 in low-income countries. The worst affected regions are in Africa and Southeast Asia, where the WHO estimate that between 66 and 76 per cent of the diarrhoeal disease burden could be prevented if access to safe wash services was provided.

The impact of this would be that almost 10 per cent of the global disease burden could be prevented by improving the water supply, sanitation, hygiene and the management of water resources.

In terms of the diseases we have already looked at, this means that 1.4 million child deaths per year from diarrhoea and 860,000 child deaths per year from malnutrition would be preventable. Two billion intestinal nematode infections – affecting one-third of the world's population – could be prevented, 25 million seriously in-capacitated people could be spared from lymphatic filariasis, and visual impairments in 5 million people with trachoma could be avoided. Two hundred million people with preventable infections from schistosomiasis could be liberated and we could stop half a million preventable deaths annually from malaria.

Dengue, Japanese encephalitis and onchocerciasis, also all linked to water resource development and management, together, cause 31,000 deaths per year worldwide and these too could be prevented.

The benefits would not only occur in terms of human health but in economics too.

The WHO study from which these figures are taken suggests that there would be a financial return of US$ 84 billion a year from the annual US$ 11.3 billion investment needed to meet the drinking water and sanitation target of the Millennium Development Goals.

The estimated economic benefits of investing in drinking water and sanitation would come in several forms: there would be health-care savings of US$ 7 billion a year for health agencies and US$ 340 million for individuals, 320 million economically productive days would be gained in the 50 to 59 age group each year, an extra 272 million attendance days a year by schoolchildren and an added 1.5 billion healthy days for children under five years of age would simultaneously achieve productivity gains of US$ 9.9 billion a year. Time savings resulting from more convenient drinking-water and sanitation services would total 20 billion working days a year, giving an annual productivity payback of some US$ 63 billion and the value of deaths averted, based on discounted future earnings, is estimated to amount to US$ 3.6 billion a year.

In terms of the respective elements of WASH, a 'safely managed drinking water service' is 'one located on premises, available when needed and free from contamination'. Additionally, an 'improved water source' refers to a range of amenities, including piped water on the premises, i.e. a piped household water connection located inside the user's dwelling, plot or yard and other sources such as public taps or standpipes, protected dug wells or springs and rainwater collection. In 2019, the UN assessed that 435 million people were still using unimproved sources for their drinking water and that 144 million were still using surface water, such as lakes and streams.

The use of groundwater is a complex subject. It has become increasingly used in places such as sub-Saharan Africa to improve access to water and achieve better resilience to climate change. These facilities are generally introduced on the assumption that the water will be suitable for drinking because of the relatively low microbiological contamination of groundwater compared to surface water. However, as we discussed in the first chapter, groundwater can be polluted by non-biological sources, such as chemicals from human agricultural and industrial production, including fluoride, arsenic or nitrates, as well as saltwater intrusion from poor water supply management. There is also, as previously mentioned, the issue of poor knowledge of many underground supplies and storage systems, as well as the longer-term viability of aquifers to be considered.

Access to sanitation services is included in target 6.2 of SDG 6, which states that by 2030, the aim is to achieve access to adequate and equitable sanitation and hygiene for all and end open defecation, paying special attention to the needs of women and girls and those in vulnerable situations. The definition of improved sanitation facilities is 'those facilities designed to hygienically separate excreta from human contact'. In 2021 over 3.6 billion people did not have safely managed sanitation in their homes, approximately 600 million people shared a toilet or latrine with other households and 494 million practised open defecation, despite huge investments and improvements in sanitation over recent decades.

According to the UN, hygiene refers to 'conditions and practices that help to maintain health and prevent the spread of diseases'. This can of course comprise a wide range of behaviours that include hand washing, menstrual hygiene and food hygiene. Hand washing with soap and water is regarded as a top priority in all settings and can consist of a sink with tap water, buckets with taps and portable basins. Despite these capacious definitions, only 14 per cent of people in sub-Saharan Africa have hand-washing facilities and globally, approximately 40 per cent of the entire human population live without the basic ability to wash their hands with soap and water at home.

Extensive WHO activity has not only assessed how many deaths could have been prevented by sufficient WASH services but which disease entities could be avoided or reduced. They looked at four main health outcomes: diarrhoeal illnesses, acute respiratory infections, malnutrition and soil-transmitted helminthiases. In 2023, they identified that 'in 2019, the use of safe WASH services could have prevented the loss of at least 1.4 million lives and 74 million disability adjusted life years (DALYs) from these four health outcomes. This represents 2.5 per cent of all deaths and 2.9 per cent of all DALYs globally. Of the measured disease entities, diarrhoeal diseases were the most significant. Over 1 million deaths and 55 million DALYs from diarrhoeal disease were linked with inadequate water facilities, and of the deaths, 564,000 were linked to unsafe sanitation practices.

Two groups merit particular mention. Children suffering from diarrhoea are more likely to become chronically underweight due to stunted growth, which in turn, makes them more susceptible to other diseases such as acute respiratory infections and malaria. The impact on women and girls introduces other considerations too. It is estimated that 263 million people worldwide spend over thirty minutes per round trip to reach an improved water source and that these are mainly women. This is time that could have been spent on their education, economic activity, culture, politics, recreation or simply rest and it represents a shocking waste of human resource in the twenty-first century. This is even before we consider female-specific hygiene and privacy needs relating to urination, menstruation, pregnancy and childbirth or the risk that many women face of violence, including sexual violence, when they must travel long distances to collect water, often on their own or in the dark. If we are serious about tackling the rights of women, we need to understand how important the issue of water is in this challenge.

So, what progress has been made since the introduction of the UN's SDGs? In March 2023, the unfortunate assessment was that 'our current trajectory towards achieving SDG 6 by 2030 is seriously off track. Without rapid, transformative progress, the world risks missing this goal, which in turn will have negative impacts across the entire 2030 Agenda for Sustainable Development.'

While hugely disappointing, important progress has nonetheless been made on several fronts. Between 2015 and 2021, the amount of water sector official development assistance (ODA) channelled through beneficiary governments increased by $100 million from 60 per cent to 72 per cent of total ODA. This shows a welcome improvement in the level of co-operation and alignment between donors and recipient governments. Across the world, there has been a welcome jump in women's participation in water and sanitation management since the beginning of the process. Seventy-eight countries, which represents 64 per cent of reporting countries, had laws or policies mentioning women's participation in 2021, compared to only thirty-five countries in 2016. In Pakistan, coverage of basic sanitation services rose to 68 per cent of the population, though the improvement rate of 2 per cent per year is not enough to achieve universal access by 2030. More positively, the number of people practising open defecation there fell by 32 per cent to 7 per cent of the population, which means that the country is on track to eliminate the practice by 2030. Next door, in Bangladesh, the government has run an extensive national public campaign on hygiene, resulting in 58 per cent of the population having access to basic hygiene services, up 16 per cent on 2015 and up 38 per cent on 2005. Bangladesh has already managed to eliminate open defecation, which was practised by 12 per cent of the population in 2005. Even in more complex areas such as the management of transboundary water, extensively covered in Part 2 of this book, there has been some good news.

The Stampriet Transboundary Aquifer System (STAS) covers almost 90,000 km² of land across central Namibia, western Botswana and the northern Cape province of South Africa, providing a groundwater resource for the 50,000 people who live there and ensuring a reliable supply for household needs, crop irrigation, livestock and tourism. Like many groundwater resources, the STAS faces challenges around overexploitation and pollution, limited knowledge of the extent of the resources and the lack of a common framework for transboundary management. A new mechanism was introduced by the governments involved to improve the understanding of groundwater dynamics, achieve better cross-border dialogue and ensure that management tools are fully shared between the governments. The STAS multi-country co-operation mechanism (MCCM) is the first for a transboundary aquifer located within a river basin organisation. It provides a beacon of hope for the future, representing as it does an innovative breakthrough for transboundary co-operation, which will hopefully be replicated elsewhere.

WATERAID

Around the world, the work of governments has been supplemented and augmented by the tireless efforts of NGOs and charities, as well as the commitment and funding of countless passionate individuals. My own personal favourite organisation is WaterAid, which has been responsible for a great deal of my own interest in the subject of water and sanitation. I first became truly aware of the human importance of the issue when, as a junior minister in the UK Foreign Office in 1996–97, I inaugurated a sanitation project in Calcutta, India. It was also where I met and became friendly with Mother Teresa, who, again, improved my knowledge and stimulated my interest in the subject.

WaterAid is an international NGO and now operates in twenty-two countries. First established by the UK water industry, it now operates as a federation with members in the USA, Japan, Australia, Canada, Sweden and India, as well as the United Kingdom. Its phenomenal fundraising activities have seen its income increase from £1 million in 1987 to £94.5 million in 2023 and it has had King Charles III as its president since 1991. Since 1981, WaterAid has supplied 29 million people with decent toilets, 28.5 million people with clean water and 28 million people with sanitation. Its reputation is such that it receives direct funding from several major international corporations as well as from non-profit organisations. Any organisation that can attract funding from as diverse a group as Boeing, the World Bank Group, Google and PepsiCo, as well as the Bill and Melinda Gates Foundation, must be doing something right.

WaterAid is a leading organisation working on WASH globally, engaging with local partners in Africa, Asia, Central America and the Pacific region to help communities establish sustainable water supplies and toilets and promote the benefits of improved sanitation and the cost of health and economic problems associated with poor hygiene practice through education. Some of its in-country successes have been spectacular. Zambia was the location of the first funded WaterAid programme in the early '80s and the organisation continues to work there today, improving connections between people and local governments to provide water solutions across the country.

In India, WaterAid also works closely with local partners to utilise low-cost technologies that deliver sustainable water, sanitation and hygiene solutions, focusing on some of the poorest communities in both rural and urban areas. Since 2010, it has evolved from being a UK-based organisation working in India to being an Indian entity, registered as Jal Seva Charitable Foundation (JSCF). The organisation remains an associate of the WaterAid Federation and has seen the transition from Sustainable Development Goals to Millennium Development Goals – in which water and sanitation have been accorded their own goal – in its work to ensure clean water, decent toilets and good hygiene are available to all.

Work in Bangladesh began in 1986, and WaterAid has now successfully collaborated with many organisations in the country to improve conditions in some of the poorest areas. The country is a good example of what can be achieved when governments and NGOs who share the same objectives co-operate. By 1996, Bangladesh was well behind its goal of achieving 'health for all' by 2000. In response, the government established community clinics (CCs), which were one-stop health centres for every 6,000 people across the country. The aim was to meet the needs of the poorest communities with facilities including family planning, free basic medicines, primary care and referral, as well as antenatal and postnatal care, all within half an hour's walk from people's homes. Following a survey which revealed a nationwide lack of WASH in CCs, WaterAid Bangladesh started work on WASH in healthcare facilities in 2016. The following year, along with a local partner, SKS Foundation, they began training community groups and those responsible for the operation of CCs to improve their skills. Again, their efforts have produced significant results with the flow of patients visiting the CCs in the areas where WaterAid intervened up by 30 per cent. Not only has an accessibility audit led to the introduction of inclusive WASH facilities but around 300 women have given birth at the CCs in the subsequent three years, compared to none before, representing a real-world improvement in women's rights and conditions.

WaterAid is a great example of how clarity of purpose, strong ethics and leadership and sound business practices can achieve great things for some of the world's most disadvantaged people. I would urge anyone who reads this book to contribute financially to their wonderful work.

THE ROLE OF FINANCE

In 2017, a position paper, prepared for a meeting of international finance ministers, by a group including the international not-for-profit organisation Water.org and the government of the Netherlands, set out, in a practical way, what they believed were some of the most pressing issues. If the international goals on water and sanitation were to be met, they said, several significant deficits needed to be addressed. These were the lack of finance for strengthening the enabling environment, the untapped use of micro and blended finance and inequities in the allocation of finance in the sector. 'These issues', they said, 'require urgent attention from Ministers of Finance as they work with relevant ministries to develop financing strategies that will enable their respective countries to meet the ambitious targets set forth in Sustainable Development Goal 6 (SDG 6).' They also presented further practical measures that needed, and still need, to be dealt with. These included the provision of infrastructure that is maintained and lasts so that people can be confident that they have regular, reliable access to water and sanitation services, that the sector becomes more attractive to much-needed additional private and public financing and that the poorest and most neglected benefit from public finance investments. Unfortunately, the Covid pandemic meant that time was lost in achieving some of these important gains, so it is now imperative that there is an unrelenting focus on all the issues that were highlighted in the paper.

THE COST OF CLEAN WATER

So, how much will it cost to give everyone access to the human right to clean water and provide the sanitation that will transform the health prospects of so many millions of people?

According to a World Bank report, published in 2017, to deliver universal, safe water and sanitation, countries will need to quadruple their spending to $150 billion a year to achieve the target by 2030. Investments will need to be targeted to ensure that services reach the most vulnerable and governments will need to engage the private sector more closely to meet the high costs, according to Guangzhe Chen, the World Bank's Vice President for Infrastructure. Admittedly, it is a higher estimate than those made by other organisations but most of us recognise from experience that major infrastructure projects seldom come in at the lower end of expectations!

This challenge comes at a time when some countries are already failing to maintain the necessary infrastructure to deal with their growing populations, never mind improving it. In 2015, for example, Nigeria provided piped water to fewer than 10 per cent of city dwellers, down from 29 per cent twenty-five years earlier. Water purity is also a growing problem. In Bangladesh, around 80 per cent of the water supply is contaminated by E. coli bacteria, meaning that tap water is more unsafe than pond water.

It is often difficult to put such large sums of money into perspective. But the prize, that almost 10 per cent of the global disease burden could be prevented, is enormous.

Some comparative figures might help. It would cost $150 billion per annum to give all our fellow human beings access to clean water and proper sanitation, in comparison to the rapidly rising revenues of the music industry, which were $26.2 billion in 2022, or the $107 billion annual global spend on cosmetics. It is a drop in the ocean compared to the global defence budget of $2,240 billion, but the figure that always impresses me is that of another liquid. This year we will spend around $1,609 billion globally on alcohol. In other words, we could supply clean water for all the people on our planet for 10 per cent of our drinks bill. Makes you think, doesn't it?

PART 4

THE FUTURE OF WATER

CHAPTER 18

WATER USE AND MISUSE

TOO MANY PEOPLE

With the total amount of water on the planet neither increasing nor decreasing and with the proportion of freshwater necessary for life restricted, any increase in demand is likely to impose increasing pressure on the availability of supply. This brings us to the thorny subject of whether we simply have too many people on our planet for its environmental safety and sustainability.

The global human population has grown from 1 billion in 1800 to 7.9 billion in 2020. UN projections indicate that this will rise to 8.6 billion by mid-2030, 9.8 billion by mid-2050 and 11.2 billion by 2100.

HUMAN POPULATION GROWTH

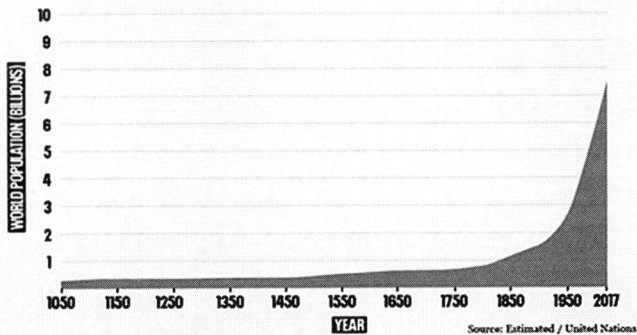

Source: Estimated / United Nations

A rising population will inevitably lead to an increased demand for food and water, and it is thought that by 2054, demand will roughly double from where it is today. Already, over 70 per cent of all existing global freshwater is withdrawn for irrigation in agriculture. Increasing dietary changes, especially the global tendency towards greater consumption of meat and livestock products, the acceleration of urbanisation and the impacts of climate change are all likely to increase water demand.

It is not just the total population that matters but how we choose to organise ourselves as societies. Big cities don't simply use more water because they have higher populations; they actually use more water per capita. While small towns (with populations of 10 to 50,000) will use between 110 and 180L of water per inhabitant per day, large cities (with populations over 250,000) will use between 150 and 300L per inhabitant per day. Why should this be? The answer lies in the kind of facilities and amenities that larger cities have to offer. A bar, for example, is likely to use between 5 and 15L per customer per day while an office will use 30 to 70L per employee per day. A hotel will use between 100 and 200L per guest per day and a commercial laundry will use a huge 100 to 200L per machine per day. If we add to this the public utilities that a city will have (a hospital will use 300 to 1,000L per bed per day), we can see how it all adds up.

To further aggravate the problem of rising populations, the demand and supply of freshwater are often mismatched with rapid population increases in areas with the highest water stress. In many parts of the world, water is already being withdrawn faster than nature's ability to replenish it. This is occurring in China, India, parts of the Middle East, Mexico, the United States, Central Asia, Africa and Spain. We will look at some specific cases and examine their wider implications later in this chapter, but if we look at where the global population exists on a continental basis we can quickly see where water stress is most likely to occur. It is virtually inevitable that the greatest crises in water will occur in Asia and Africa as their rising populations and increased urbanisation clash head-on with limited water resources.

THE CONCEPT OF 'VIRTUAL' WATER

While these high-population, low-water-resource areas may be at the forefront of the twenty-first century's water problems, all of us, as global consumers, are responsible for the use of water far from where we may live.

The total water demand of any country is usually measured by the amount of water withdrawn from rivers, reservoirs, aquifers and lakes for its consumption and economic production. The problem with this measure is that it does not consider the amount of water consumed via imported products and so can give a false impression of the impact that the country's domestic human and economic activity has on the wider global water resource.

To help resolve this problem of measurement, Tony Allan, Professor of Geography at King's College, London, introduced the concept of 'virtual water' in 1993, to demonstrate how water resources in one country are used to support consumption in another. Virtual water is defined as the hidden flow of water in food or other commodities that are traded from one place to another. When goods and services are exchanged, so, therefore, is virtual water. Although virtual water may be invisible to the end-user of a product or service, water has been used throughout the value chain and is thus part of the product's inherent value, both economically and environmentally.

The expansion of international trade means that many countries that have scarce supplies of water have become more dependent on the water resources of other countries to provide what their consumers want and need. For example, as we will see, water-scarce countries like Saudi Arabia, who used to attempt to grow their own animal feed, now import it from countries with much higher water availability. These countries will have entered commercial contracts with Saudi Arabia to ensure that their plentiful water supply provides the country with what it needs in terms of animal feed, while not overly stressing the limited water supply that Saudi Arabia has. This is a good example of where balanced trade can be a winner for both sides. Some countries with limited water supplies, however, may still be tempted to export goods, which will exacerbate their domestic water shortage as they seek greater development and income from exports. The concept of virtual water is important because, in Dr Allan's words, 'it enables us to understand why we enjoy the illusion of water and food security despite strong evidence that the water resources available to sustain our national economies are inadequate.'

Central to the understanding of both virtual water and a similar, but distinct, idea known as water footprint, is the division of water supply into 'blue water' and 'green water'.

Blue water is the freshwater that flows naturally at the land surface, such as in rivers or in underground storage systems like aquifers. It can be easily engineered and pumped and is therefore available for all sorts of economic activity, including household consumption, farming and energy generation. When it is unregulated it can produce a water overdraft, where consumption is greater than its supply or replenishment, a concept that many of us may understand from our bank accounts, but, in these cases, one which can take hundreds if not thousands of years to rebalance.

Green water exists, as its name suggests, in the soils of farmland and natural landscapes. Unlike blue water, it cannot be engineered and pumped, and it can only be moved as water embedded in food supply chains. In fact, the only pumping done is by nature in the water cycle. In other words, the competition for this water is between those farmers trying to irrigate their crops and the forces of nature itself. As such, it is arguable that green water is much better able to look after itself than blue water. It is worth noting that a given volume of green water will produce the same amount of food as the equivalent volume of blue water which immediately makes clear the need to utilise the use of green water for food production in the appropriate parts of the world while avoiding the use of blue water to produce goods in an inappropriate climate.

The United States, the world's largest economy, is the largest exporter of water through trade. It is also (because of the volume of its trade) the largest importer but, as this only offsets around 75 per cent of the water exported, it remains one of the countries with the largest water trade deficits. China, India and Brazil are also large exporters, but as water pricing to producers tends to be independent of the availability or scarcity of supply, there is no market mechanism to slow the use of unsustainable water resources. Argentina, Canada and Australia (despite the latter's growing water problems) are also part of the group of major net water exporters, while Japan, European countries, North African and Middle Eastern countries and Mexico are the world's biggest net importers.

WATER FOOTPRINT: HOW MUCH WE REALLY USE

Water footprint is a slightly different concept from virtual water, but it is another useful tool in helping us to understand how water consumption can be more complex than it might initially seem. It shows the extent of water use in relation to consumption by people and is expressed in several different ways.

It can be viewed as the water footprint for products, for companies, for individuals or for countries. It also adds a new category of water to the blue and green water mentioned previously, that is, grey water.

Grey water consumption is the volume of water needed to dilute pollutants (whether produced by industrial discharge, agricultural runoff or untreated municipal wastewater) to meet the necessary water quality standards. The more a country or a company pollutes, the more grey water it uses or, thinking of it another way, 'owes' to the global water system.

The largest water footprints belong to the United States, India and China. Water pollution accounts for about 30 per cent of China's footprint, which is not surprising when China accounts for around 25 per cent of the entire world's water pollution.

A product's water footprint is the total volume of freshwater used through all the stages of the production chain up to the finished item. It is, therefore, not only measured by the total volume of water required to produce the item but where and when the water is utilised in the process.

It will come as a great surprise to many just how much water is required to produce many of the items that we take for granted in our diets. It is also easy to see the impact that changing diets have on global water demand. Since it takes over 15,000L of water to produce 1 kg of beef but only 4,300L of water to produce 1 kg of chicken, it is easy to see how a shift in the dietary tastes of, say, the 1.4 billion Chinese population from chicken to beef could have a potentially catastrophic impact on water consumption. Again, given that vegetable production requires a little over 300L per kg it is easy to see why increasing urbanisation and the shift from vegetarian to meat diets can have a massive impact on water balance and availability.

Below, I include examples of the amount of water required for many of the foods that are part of our regular diets.

Before everyone piles in on meat eaters, there are other, more luxurious, items that have even higher water footprints. Bovine leather requires 17,093L of water to produce 1 kg, chocolate requires 17,196L of water to produce 1 kg and vanilla beans, a staggering 126,505L of water to produce 1 kg. It is something to ponder when we have our next ice cream!

Of course, food production is not the only source of agricultural water consumption. Textile production also requires a great deal of water and around 70 per cent of the water footprint for the fibre production used in the entire textile industry is due to a single crop: cotton. One kilogram of cotton requires between 10,000 and 20,000L of water for its production. This variation depends on where in the world it is grown. Around 73 per cent of all global cotton production requires irrigation, i.e., blue water consumption, and is grown in countries with warm climates where freshwater supply is already stressed. Almost 70 per cent of total global cotton production takes place in China, India, Pakistan, Uzbekistan and the USA. It is estimated that the amount of water required to irrigate global cotton crops is equivalent to twice the total annual water footprint for the whole United Kingdom (i.e., for our domestic, industrial, farming and water-cleansing needs).

When we consider individual consumption, and measure the footprint per capita, we can see rather different patterns from those we will see by looking only at the level of a particular nation. The United States has one of the highest human consumption levels of around 2,800 m^3 per year per person, compared to a global average of 1,400 m^3 per year per person.

Two factors affect per capita water footprints – the pattern of the goods being consumed and the water intensity of the goods being consumed. We have already mentioned the amount of water that is required to produce beef so it is not surprising that the US per capita figure should be so high when we consider that the average American eats 4.5 times the global average of beef each year – around 95 pounds. That is a lot of burgers!

The UK, as one of the world's biggest economies, imports a wide range of products including large quantities of livestock products, tea, sugar and cereals, as well as chemicals, machinery and other manufactured goods.

According to World Wildlife Fund (WWF) calculations, this equates to around 4,645 litres per person per day, the equivalent of seventy-five cycles by a standard washing machine. The fund breaks this figure down as 'agricultural products, 3,400 litres per person per day, with cotton alone representing 211 litres per person per day; industrial products, 1,095 litres per person per day; and household water, 150 litres per person per day.' These figures include both the internal footprint, the water used from inside the UK to grow the food consumed in the UK, and the external footprint, the water used to produce the goods that we import.

This all adds up to around 1,300 m^3 per person per year, just below the global average, but is surprisingly large for a country that experiences such a huge amount of rainfall. It is indicative of both the high level of imports coming into the country (the UK regularly runs a very large trade deficit) and a lack of efficiency in the use of domestic blue water.

Putting it another way, 62 per cent of the UK's total water footprint is accounted for by the use of water in other countries, which means that the food and clothing we consume is inextricably linked to the management of water sources elsewhere in the world, whether we are aware of it or not.

More than a third of all the products that make up the UK's external water footprint originate from Brazil (soybeans, coffee and livestock products), France (mainly seasonal produce), Ireland (meat products) and India (cotton, rice and tea). Ghana is also a major exporter of cocoa, but as this is mainly rain-fed, the footprint for this is lower.

To make real sense of the environmental impact that water footprint has, we must consider not only the product being produced but also where that is happening. For example, the 140L of water required to produce coffee and deliver one cup might be of no harm to water resources if its cultivation occurs mainly in humid areas, while it could be enormously damaging in more inappropriate, arid regions. While we need to interpret the information carefully by considering other factors such as geology, topography, population and demographics, high-water footprint calculations reasonably suggest that genuine environmental concern is appropriate.

WHAT CAN BE DONE

To start with, we must understand that the water we use as individuals is not just what we eat, drink or wash with, but the water that is contained in all the goods that we use as part of their production, wherever that occurs in the world. We need to become more aware of the importance of over-abstraction and pollution in our water systems, especially if we are to protect our current biodiversity, and that awareness needs to extend globally and not just to our own locality or nation. We also need to develop and encourage clear priorities for those who devise national and international political policy and for those who manage the water systems. The WWF have set out three priorities to achieve these goals. They are: to provide enough water for the basic needs of local people, to ensure that sufficient water is left in rivers, lakes and aquifers to support essential ecosystems and to share water fairly between different users and encourage greater efficiency. I don't think that any of us could disagree with these basic aims. When it comes to good agricultural practices, we need to use our limited resources wisely to improve the quality and productivity of soil so that it retains more moisture and we can use more green water and less blue water. Recent advances in agricultural technology, such as hydroponic, aquaponic and vertical farming have shown that it is possible to massively improve efficient growth while minimising water usage. We need to find ways to encourage, develop and disseminate such technology.

One final thought is that we can all diminish our water footprint by reducing food wastage. It is estimated that around 40 per cent of the food in the United States is never eaten, food which will have required a great deal of water to produce it. The old-fashioned idea of not leaving food on the plate (admittedly more of a legacy in those countries who experienced rationing in the Second World War) would be easier if we didn't overfill oversized plates in the first place! We should all become more conscious of how we use our own food, how we use leftovers and of not overstocking our fridge every week. There are, after all, a plethora of TV cooking programmes to give us plenty of tips. A change in our behaviour in relation to food waste may, as with many changes in our behaviour, only make a small contribution to our overuse of global water but if we all do the same then we might just make a difference.

A NEW CONCEPT: COMPARATIVE
ECOLOGICAL ADVANTAGE

In his great book, *The Wealth of Nations*, the Scottish economist Adam Smith gave us the concept of absolute advantage in trade which suggests that countries should focus on the goods and services they can produce themselves and then use trade to acquire the products that they cannot make. The concept was explored and elaborated upon by economists like David Ricardo to produce the theory of comparative advantage. This idea refers to the ability to produce goods and services at a lower opportunity cost compared to the competition. For example, the southwest of England, where I live, has a comparative advantage in its dairy industry and is able to produce abundant quantities of cheese and milk as a result of its relatively wet climate and the amount of land available for dairy cows. Less developed countries, on the other hand, may have a relative advantage in the production of consumer goods that require intense labour as a result of their lower labour costs. Where does the environment fit into this narrative? Perhaps it is time to introduce a new concept – that of Comparative Ecological Advantage (CEA). This would suggest that it is both right and responsible for countries to focus on producing those goods for which it has a natural ecological advantage, such as an abundant water supply for producing particular crops. In addition, responsible consumers should seek to source products from countries where production does not result in severe environmental degradation. Again, taking cotton as an example, countries who have limited water resources should not produce crops that will worsen natural limitations and deplete scarce resources, and those purchasing should try to increase their awareness of where their products come from. As consumers we should look for foods, such as fruit, from those countries whose natural conditions and water supplies are more compatible with sustainable production rather than those where we may, completely inadvertently, be worsening an already difficult situation. The concept of CEA is one which should be developed in the area where trade and environment intersect in a Venn diagram.

It has been said that however far away we may be from it physically, as consumers we are all at the water's edge. The choices we make about the foods we eat, the clothes we wear and the objects we choose to own and use will influence the amount of water that is used somewhere in the world. Ultimately, however far away, it is still the same water.

CHAPTER 19

CLIMATE CHANGE AND HUMAN HEALTH

CLIMATE EXTREMES: FIRE AND RAIN

By the very nature of its complexity, the number of variables involved and the constantly changing database, is difficult to precisely predict the effects that climate change is having, or might have in the future, on human health.

A much easier task is to make clear the predictable risks that we might have to face and establish from these the kinds of health challenges that might arise.

The first risk comes from the effect that changing patterns of flood and drought might bring in their wake, especially when linked to more frequent extreme weather patterns. The second risk is the ecological disruption that may impact the availability of drinking water, food security and changes in the pattern of vector-borne disease. The third risk comes from the consequences of social dislocation and population movement, primarily on the groups themselves but also on any new host populations that they encounter.

The intensity and frequency of extreme rainfall and drought have both increased markedly over the past twenty years and have progressively resulted in crop damage or failure, damaged infrastructure and even major humanitarian crises.

Technology is increasingly being harnessed to improve our knowledge of the world's water patterns. A pair of satellites known as GRACE (Gravity Recovery and Climate Experiment) have helped us to measure changes in the earth's total water storage and the sum of all the water on and in the land, including groundwater, surface water, ice and snow. Using cutting-edge technology that identified where land was much drier or wetter than usual, scientists looked at over a thousand events from 2002–21.

In terms of intense droughts, they noted the record-breaking conditions in northeastern South America from 2015–16 and the ongoing drought in the southwestern United States that has brought Lake Mead and Lake Powell, two of the nation's biggest reservoirs, to worryingly low levels.

When they looked at unusually heavy rainfall, they discovered that the most extreme rains were happening in sub-Saharan Africa, with extreme rainfall also occurring in central and eastern North America and in Australia from 2011–12.

In 2023 there were several spectacularly heavy rainfall events around the globe. In September, streets in Libya, Spain and Greece were turned to rivers by torrential downpours, while widespread flooding hit Hong Kong and New York City. In India, unusually heavy monsoon rains claimed many lives as mudslides swept through the Himalayas, while Zagora, in Greece, saw a new record rainfall of 30 inches in twenty-four hours, the equivalent of eighteen months' normal precipitation.

This is a practical example of the combined effects of climate change and the intrinsic properties of water. As we have seen, a warmer atmosphere produces more precipitation events, and as the temperature rises, the warmer atmosphere can hold more water vapour. In fact, the atmosphere can hold about 7 per cent more water vapour for every 1°C (1.8°F) rise in atmospheric temperature. As the evaporation of water from the oceans and the land both increase due to higher temperatures, this larger volume of water must eventually come back down to earth somewhere or other.

When it comes to the countries most affected by increased flooding, in terms of the absolute number of people exposed to flooding per year, the top five are all in Asia: Bangladesh, India, China, Vietnam and Cambodia. When viewed as the percentage of the population who are at risk, Cambodia is then top, followed by Bangladesh, Vietnam, Bhutan and India. Apart from the obvious consequences of physical injury and drowning, more frequent and severe floods will produce other problems, such as increases in the incidence of infectious diseases like hepatitis, the diarrhoeal diseases and vector-borne diseases. As well as the greater risk of exposure to toxic substances from agricultural or industrial discharge into the water, contamination of the because of flood waters containing chemicals and nutrients can have a major impact on the quality of drinking water, as can saltwater intrusion.

In cooler climates, such as those at higher altitudes, the effect of flooding because of climate change is slightly different. Normally, snow melts in the late spring or summer and feeds the rivers downstream with water. With warmer winters now occurring, however, there is less snowfall and more rain. This can have a profound effect on the risk of flooding as, while snowmelt is usually a gradual process, rain creates runoff that reaches the rivers much more quickly. As a result, floods precipitated by rain can be much larger than those caused by snowmelt alone and may also intensify another potentially deadly risk, that of mudslides.

It is thought that the risk of mudslides is also increased because temperature rises induced by climate change increases the risk of wildfires. According to Frances Davenport of Colorado State University, areas that have been burned in wildfires are more susceptible to mudslides and debris flows during extreme rain, both because of the lack of vegetation and changes to the soil caused by fire itself. Changes to the soil due to fire can make it more impermeable to rain so that more water will end up in streams and rivers, thus creating worse flood conditions. Pictures of wildfires across the globe have become one of the most totemic symbols of climate change and when these are combined with increasingly extreme rain events, it is easy to see why flooding and wildfires could become such devastating partners in destruction.

CLIMATE CHANGE: DUST AND DROUGHT

As 2023 has progressed, anomalously warm conditions have continued to occur across all continents. It was the warmest October on record for Asia and South America, with Africa, Europe and North America having the second warmest October on record.

According to the global drought information system, dry conditions persisted across much of Canada, South America, Australia, northern China and the Mediterranean region during October 2023. It is just a snapshot, and it should be noted that welcome rains fell across some of the drought areas on other continents. Yet, it is a snapshot that is an increasingly apparent pattern.

Drought has reduced crop yields and raised the price of foodstuffs worldwide. The GEOGLAM crop monitor, created to improve food security by monitoring agricultural conditions and the outlooks for production at national, regional and global levels, has indicated that agriculture has been most threatened in parts of the Americas, east Africa, eastern Europe, southern Asia and parts of Australia and Indonesia. The Famine Early Warning System Network (FEWS NET), created by the United States Agency for International Development (USAID) and whose reporting focuses on acute food insecurity, has revealed continuous problems in parts of central and south America, Southwest Asia and much of Africa. In Europe, where a record number of heatwave spots created numerous wildfires and caused chaos across the Mediterranean tourist industry, extreme heat, wildfires and drought have decimated much of the world's olive oil harvest yet again, driving prices to a record high of US$9,000 per metric ton. Spain, the source of half the world's olive oil supply and the global price setter, reported in May 2023, a drop in production of 48 per cent compared to the previous year. Insurance claims paid out to winegrowers as a result of drought were estimated at €100.5 million in 2023, which was more than double the payout for wine grape damage in 2022. It is a pattern repeated in other agricultural sectors.

Turkey produces 70 per cent of the world's hazelnuts and it has been battling drought and damage to crops by insects since May 2023, affecting the prime growing season. Studies suggest that not only have heat and drought reduced wheat and maize yields by up to 40 per cent worldwide but, shockingly, projections suggest that for every degree Celsius (C) rise in temperature, there would be a 6 per cent loss in global wheat yields.

This all provides a worrying outlook for the rise in global food insecurity and its knock-on effects for human health. In Africa alone, over 23.5 million people are currently suffering high levels of acute food insecurity related to drought in Ethiopia, Kenya and Somalia.

Insufficient nutrient intake can increase vulnerability to infection, especially among groups such as children under five years old, pregnant and lactating women, the elderly and those suffering from chronic infections such as tuberculosis (TB) or HIV.

The diarrhoeal diseases we have discussed are also likely to rise because of limited access to water supplies, poor hygiene and inadequate sanitation.

Significant population displacement following drought with time spent in overcrowded, communal emergency shelters, coupled with poor hygiene, can lead to an increased incidence of respiratory illnesses or skin diseases.

It is a sad picture which is likely to become ever sadder as extreme weather patterns take an increasing toll on the most vulnerable populations. These ever-more frequent climate-induced events can also be exacerbated by recurrent natural trends. The longest drought ever recorded lasted 172 months (14.33 years) in Arica, Chile between 1903 and 1918. This should not perhaps be surprising given how much the climate there is affected by the El Niño and La Niña climatic events.

EL NIÑO AND LA NIÑA

The El Niño-Southern Oscillation (ENSO) is the most influential natural climate pattern on earth, and it moves between three phases – neutral, La Niña and El Niño. Generally, El Niño occurs more frequently than La Niña.

The oscillation produces a variation in winds and sea surface temperatures over the eastern Pacific Ocean, which affects the climate of large parts of the tropics and sub-tropics. Normally, relatively cold water from the Southern Ocean moves up along South America's west coast, where it is reinforced by the upwelling of more cold water along the coast of Peru. As this cold water reaches the equator, the trade winds drive it on a path westwards, where it is slowly heated by the sun so that the sea surface temperature (SST) is around 10°C warmer in the western Pacific than in the eastern Pacific.

During the warmer El Niño years, this normal pattern is interrupted and the cold water weakens or disappears so that the water in the central and eastern Pacific becomes as warm as the western Pacific. El Niño is accompanied by high air pressure in the western Pacific and low air pressure in the eastern Pacific which produces above-average levels of rainfall over the central and eastern Pacific and below-average levels over Indonesia, India and northern Australia. At the same time, the trade winds, which normally blow from east to west along the equator, either weaken or start blowing in the other direction.

In the cooler phase of ENSO, La Niña, the sea surface temperature in the eastern Pacific falls to below-average levels, with high air pressure in the eastern Pacific and low pressure in the western Pacific, the opposite of that found in El Niño periods.

These events are not regular but usually occur two to four times per decade, with most events lasting about a year, reaching their intensity late in the calendar year.

The neutral phase indicates that conditions are near their long-term average.

The El Niño phenomenon is not a new one and the increased rainfall caused by El Niño has been found in coral specimens that are at least 13,000 years old. It is even believed that a strong El Niño effect, between 1789 and 1793, contributed to extremely poor crop yields in Europe and produced shortages that helped trigger the French Revolution, while the extreme El Niño of 1876–77 is thought to have produced one of the deadliest famines of the nineteenth century, with up to 13 million people dying in China in 1876 alone.

The weather pattern effects of El Niño and La Niña are, in many ways, mirror images of each other. In western Asia, the El Niño phase is associated with increased precipitation, whereas with La Niña, it is reduced. El Niño is associated with warm and wet weather along the coast of northern Peru and Ecuador, along with major flooding, especially when the event is strong. By contrast, La Niña usually produces heavy rains over Malaysia, the Philippines and Indonesia. With El Niño, snowfall is greater than average across the southern Rockies and Sierra Nevada mountains, whereas during La Niña snowfall is above average across the Pacific northwest and the western Great Lakes. El Niño years generally result in less active hurricane seasons in the Atlantic Ocean but an increase in tropical cyclone activity in the Pacific Ocean. La Niña years tend to see above-average hurricane development in the Atlantic and less development in the Pacific basin.

Studies have suggested that climate change has doubled the likelihood of strong El Niño events and increased the likelihood of powerful La Niña events ninefold.

These phenomena can have major effects on food and water security for populations around the globe with huge variations from the normal climatic conditions producing either increased rainfall, causing floods or decreased rainfall, causing drought.

Any acceleration or increased intensity of the oscillation induced by climate change can have calamitous implications for food insecurity and human health. They are also likely to have an impact on vector-borne diseases, including one of the world's biggest killers, malaria.

MALARIA

The WHO estimates that there were 608,000 malaria deaths in 2022. Tragically, over three-quarters of all deaths from malaria in sub-Saharan Africa were among children under five. The number of global cases continues to rise, with around 233 million cases in 2019, rising to 249 million cases in 2022. The relationship between climate and malaria is a complex one, as a complicated interplay between humans, mosquitoes and the Plasmodium parasite which causes the disease already exists.

Humans can become infected with malaria when they are bitten by the female Anopheles mosquito if the mosquito itself is already infected by Plasmodium parasites, which it then passes on to its new host. Two Plasmodium species, P. falciparum and P. vivax, are the most dangerous, with P. falciparum malaria able to progress to severe illness and even death within twenty-four to forty-eight hours if left untreated. The most prevalent form of the parasite on the African continent is P. falciparum, while P. vivax is more dominant in countries beyond sub-Saharan Africa. Many people will be surprised to discover that, until 1975, malaria was endemic in Europe. It was eliminated (notwithstanding the increase in global temperature) due to improved water and farm management, improved healthcare and generally better socio-economic conditions.

In general terms, reducing malaria rates is best achieved by avoiding mosquito bites, controlling mosquito populations, using preventative medicines, vaccines and early medical intervention.

Simple measures to reduce being bitten include sleeping under mosquito nets in areas where the disease is prevalent or installing window screens, using repellents (especially after dusk) and wearing full-length protective clothing (i.e., not shorts and T-shirts after dark).

Controlling and eliminating mosquitoes is a key part of controlling infection with the two core interventions being insecticide-treated nets (ITNs) and indoor residual spraying (IRS). There is, of course, a constant battle to keep the equipment and supplies up to date in often poor and widely dispersed populations, a huge task for the many governmental and NGO groups engaged in the task. Another problem is the change in mosquito behaviour as humans are increasingly finding themselves being bitten before they go to bed, with the insects then resting outdoors, evading exposure to insecticides.

Those intending to travel to malaria-endemic areas should consult a doctor well before they travel, as some of the drugs used to prevent the disease need to be started two to three weeks before departure. The medication must be taken not only during the stay in an infected area but for four weeks afterwards, as parasites, which may have found their way into the liver, can still emerge in the bloodstream during this time.

Populations who are at higher risk of malaria can be treated with preventive medicines at key periods, whether or not they show signs of infection. These can be administered to entire populations at particular times (mass drug administration) or aimed at specific groups, such as pregnant women and school-age children, and they represent safe and cost-effective strategies that complement other key anti-malaria strategies.

As of October 2023, the WHO recommends a programme of malaria vaccines for the prevention of P. falciparum malaria in children living in malaria-endemic areas, prioritising those areas of moderate and high transmission. This applies to both RTS,S/AS01 and R21/Matrix-M vaccines. The first malaria vaccine, RTS,S, was recommended by the WHO to prevent malaria in children in 2021. It has reached nearly 2 million children in Ghana, Kenya and Malawi through the malaria vaccine implementation programme (MVIP) since 2019. The newer R21/Matrix-M vaccine marks the culmination of thirty years of research at the University of Oxford's Jenner Institute and was jointly developed by the university and the Serum Institute of India. Vaccination has already been shown to significantly reduce malaria, and severe deadly malaria, among young children and is a major medical milestone, which represents significant hope for the future.

Malaria infection is serious and always requires treatment with medicine, although early diagnosis and treatment can reduce its severity, prevent deaths and help to reduce transmission. The choice of medical treatment will depend on the type of malaria, indications of resistance, the patient's age and weight and whether the patient is pregnant.

Symptoms of malaria usually begin ten to fifteen days after being bitten by an infected mosquito. Since symptoms are generally non-specific, with fever and headache, it is important to have a high index of suspicion, especially if there has been travel in or through endemic areas. They can also be mild for those who have had malaria before, so any suspected cases require early testing. Early treatment for mild malaria can stop the infection from becoming severe or life-threatening. This is particularly important for those who might be at greater risk, such as children under five years old, pregnant women or those with HIV or AIDS. Malaria infection while pregnant causes an increased risk of premature delivery or delivery of a baby with a low birth weight.

In untreated or severe cases, worsening symptoms can include extreme fatigue, dark or bloody urine, jaundice, abnormal bleeding and even convulsions, impaired consciousness and breathing difficulties.

No anti-malarial drug offers 100 per cent protection and must be combined with personal protective measures. Anyone travelling to an area where there may be endemic malaria should check with public health information sites as the drug of choice will vary from country to country and patient to patient. Chloroquine is the preferred treatment if the parasites are sensitive to the drug but in many parts of the world, growing resistance means that this is no longer an effective treatment. Increasingly, Artemisinin-based combination therapies (ACTs), which contain two or more drugs that work against the malaria parasite in different ways, are being used. Atovaquone/Proguanil (Malarone) is useful for last-minute travellers as the drug can be started one to two days before travelling to an endemic area and only has to be taken for seven days after travelling rather than the full four weeks. Doxycycline, a commonly used antibiotic, can also be used for malaria prophylaxis, and although it can be started one to two days before travel, it needs to be taken for four full weeks afterwards. However, as it can prevent some other infections, such as leptospirosis or Rickettsia, it can also be a useful choice for those who will be wading through, or swimming in, freshwater. If in doubt about the most appropriate medicine to take, consult a doctor!

WILL MALARIA GET WORSE WITH CLIMATE CHANGE?

The scientific world is somewhat divided on how climate change will impact patterns of malarial disease, with some people suggesting that there will be globally elevated transmission, with others believing that the repercussions will be limited. There is more agreement that other mosquito-borne diseases, such as dengue fever, may pose a greater risk of wider transmission. What can be agreed is that the factors that dictate the spread of malaria, such as temperature, humidity, rainfall and other climatic conditions, can impact the lifespan of the mosquito and the development of the malaria parasites inside them.

Current data seems to imply that the higher temperatures, humidity and rainfall resulting from climate change are causing mosquitoes to be present in much greater numbers at higher altitudes, which carries the risk of malaria's wider geographical distribution. At lower altitudes, where the disease is already more prevalent, increased temperatures may affect the parasite's growth cycle, causing the mosquito to develop malaria faster and thereby increasing transmission rates. It is not, however, only increased rainfall that produces an increased risk of the disease. Drought, and the accompanying higher temperatures, can cause flowing rivers or lakes to dry up into smaller and warmer stagnant pools which then provide the perfect breeding ground for mosquitoes, so we must consider the potential impact of climate change in the round.

The effect is well illustrated by the impact of the El Niño phenomenon on the incidence of malaria. In Colombia and Venezuela, malaria cases increased by more than a third following the dry conditions associated with El Niño, while in Sri Lanka, the risk of the disease increased threefold following the failure of monsoons associated with El Niño. Where rains have increased, in South Africa and western and north-western India for example, there have also been malaria epidemics. To further complicate the picture, very heavy rainfall can wash out breeding sites and produce some reductions in the disease, so there is at least some element of swings and roundabouts.

One of the main worries attached to the impact of the changing climate is that it may reintroduce or increase the spread of malaria in tropical or temperate countries that have already eliminated or controlled transmission. The fear is that such countries could be prone to epidemics since surveillance and measures to control malaria may not be as intense or effective as when the disease was a major public health problem. A World Bank report has suggested that by 2050, climate change could result in parts of China, South America and sub-Saharan Africa, which are currently malaria free, becoming at risk of infection.

DENGUE FEVER

Dengue fever is another mosquito-borne illness affecting tropical and sub-tropical areas, most commonly Southeast Asia, the western Pacific islands, Latin America and Africa. Recently, however, the disease has been found in new areas with outbreaks occurring in Europe and southern parts of the United States. The disease is caused by any one of four types of dengue viruses, with most people experiencing no signs or symptoms. When they do occur, they usually begin around four to ten days after an infected bite and produce a high fever, headache, nausea and vomiting, muscle and joint pain, swollen glands and a rash. Severe infections can be life-threatening, with the number of platelets in the blood, which are required for clotting, dropping quickly and with potential internal bleeding, organ failure or even death. Severe symptoms include bleeding from the gums or nose, blood in the urine, stools or vomit or unexplained bruising. Patients who develop breathing difficulties, in particular, require rapid medical intervention. Women who develop the disease during pregnancy may also spread the virus to the baby during childbirth.

Those who have been infected with dengue fever will have a natural immunity to the particular virus that caused the illness, but they will still be susceptible to any one of the other three viruses at a future point. A vaccine is available, but only for those who have previously had dengue fever or a blood test that shows previous infection, even without symptoms. Unfortunately, for those who have not had dengue fever before, receiving the vaccine appears to increase the risk of severe dengue fever in the future. Other preventative measures are the same as for other mosquito-borne diseases: staying in well-screened or air-conditioned premises, wearing protective clothing and using mosquito nets and repellent. It is thought that climate change is behind the disease's huge increase in places like India. In 2022, 230,000 infections and 303 deaths were attributed to dengue fever in a country where severe disease was rare before 1996, but where the incidence has dramatically increased over the past decade. The organisation and lack of mosquito control programmes are likely to be part of the explanation, but climate change has increased both the temperature and frequency of heavy rainfalls, which are conducive to mosquitoes expanding their geographical range. Next door to India, in Pakistan and Nepal, there have been similar increases in incidence, with countries much further afield, such as Brazil and Niger, witnessing rapid increases in infections. Even in Europe, the virus can now be found in two-thirds of France where no locally transmitted cases had been recorded before 2010 but where sixty-five cases were reported in the first nine months of 2022 alone. It is now conjectured that the mosquito that transmits dengue fever is present in twenty-two European countries. This is a situation that we can expect to get worse in the future.

CHAPTER 20

DIRTY DIRTY WATERS

HUMAN POLLUTION: OUR CRIME AGAINST NATURE

Of all the issues that we have looked at relating to water, perhaps the most depressing is the issue of pollution, a result of our own careless and destructive behaviour. Even though the surface waters of our planet – our rivers, lakes, seas and oceans – are necessary, not only for human life but that of other mammals, amphibians, reptiles and fishes, human behaviour and the increased size of the human population have created new dangers for global ecosystems. Sewage discharge from human megacities, toxic waste from industry and ever-greater amounts of agricultural pollution have poisoned our most precious resources. How do we stop this destructive behaviour before it's too late, and how can we remedy the damage we have already done?

In the UK, agriculture is the single largest cause of river pollution, contributing around 40 per cent of the problem and runoff, where fertilisers and animal manure are washed into rivers, is a major contributor. Sewage discharge, industrial waste and careless and senseless human litter all contribute further to the problem. Excessive amounts of nitrogen and phosphorus in our rivers not only affect water quality, but aquatic ecosystems and human health. Fertilisers and manure which are rich in nitrates and phosphates often leach from fields into nearby rivers and intensive factory farms result in massive amounts of animal manure that regularly finds its way into both rivers and groundwater. As discussed earlier, excessive amounts of these nutrients can produce the phenomenon of eutrophication, which is characterised by the rapid growth of algae which can block sunlight, reduce oxygen levels in the water and produce dead zones where aquatic life struggles to survive. Pesticides, which also enter our rivers through runoff and seepage, can harm fish and other aquatic life, which not only leads to their declining populations but has an effect all the way up the food chain to birds and mammals who will also suffer as a consequence.

Pollution with several specific items, including antibiotics and hormonal products, has rightly attracted a growing amount of attention recently, as the public has become more aware of the environmental and health implications of our own behaviour.

Antibiotics can enter rivers via runoff and groundwater or by leaching from dumping, often illegally, in landfill sites. Intensive animal farming is a prime cause of the problem with preventative antibiotics being used to limit infection from animals that are kept in highly cramped and potentially disease-ridden conditions. Antibiotics in our rivers can produce antibiotic resistance in aquatic life and livestock and a recent UK study has found superbugs like MRSA in the UK's waters, especially in areas where intensive farming is most concentrated.

Recently, hormonal products have also come under increased scrutiny. A range of natural and synthetic oestrogen hormones are now widespread as environmental contaminants. The cattle industry, which uses growth-regulating steroids to improve capital growth rates, is one of the major sources of oestrogens released into the water. Sewage release, with human oestrogens excreted through urine, is another source. Pregnant women excrete a range of oestrogens and although oestrogens from birth-control pills have been shown to contribute to the feminisation of aquatic species, this has been shown to have less effect on the total oestrogen content of drinking water than other sources. Lately, there has been a great deal of coverage on the subject, often on social media and most of it scientifically unfounded. However, further detailed studies are required because the relationship between oestrogens and adverse reproductive outcomes is not fully understood and the risks need to be considered alongside exposure from other sources.

RIVERS OF SHAME

The shameful list of the world's most polluted rivers is a long one and it affects all the continents. Over half of Asia's rivers, 25 per cent of those in Latin America and up to 25 per cent of those in Africa are affected by some sort of serious pollution. Inevitably, this has the most adverse effect on the poorest populations who often obtain their drinking or cooking water directly from surface water sources such as rivers.

The Ganges, India's most sacred river, has the dubious distinction of being the world's most polluted major river, with one of its tributaries, the Yamuna, recognised as the most polluted urban river on the planet. Such is its dreadful state of pollution, its water is anoxic (with zero oxygen), full of faeces and unable to support any kind of aquatic, plant or birdlife until it gets 200 km downstream where it gradually starts to recover. As well as being a floating waste dump over which animals and human bodies are incinerated, the river has a layer of floating plastics and other waste to add to its horrific brew.

Another candidate for the toxic crown is the River Salween in Southeast Asia. As it flows from Tibet, through China and Burma, it picks up waste from industrial parks, causing it to contain frightening levels of sulphur, lead, mercury, cadmium, copper and zinc. It is reputed that the river is now so poisonous that several former fishing communities have now turned to selling the glass and plastic that they pick out of the river instead.

Rivers in every part of the world are affected by this hideous trend. In Europe, the Sarno River that flows through Italy is one of the most polluted and a real tragedy because the clean water at its source, suitable for drinking, is gradually being polluted by industry and agriculture throughout its course, to the extent that there has been an alarming increase in cases of liver cancer in the area. The river's tendency to flood and create mudslides means that this toxicity, alongside food waste treatment, can find its way into the earth.

Another European river, the Danube, has high levels of chemical, particularly pharmaceutical, waste, and extremely high levels of farming pesticides. Added to these hazards, there has recently been increasing scrutiny of the role that Serbia plays in the pollution of Europe's second-longest river. Around a third of Belgrade, Serbia's capital city of 1.6 million, has no connection to drainage systems and relies instead on septic tanks that trucks empty straight into the river. The rest of the population spew their unprocessed waste into the river through around 100 sewage drains. It has been estimated that around 190 million m³ of wastewater – the equivalent volume of 60,000 Olympic-sized swimming pools – is annually poured into the river. Tourists who enjoy the plentiful cruises down the much cleaner parts of the river, in Vienna or Budapest, might want to reflect not only on the state of the river itself but that the high level of river traffic is also contributing to the river's contamination.

The United States has at least three contenders on the toxic list:

the Rio Grande, the Mississippi and the Passaic River in New Jersey. The Rio Grande flows through a region largely prone to drought and has the disadvantage of the proliferation of invasive species that consume a lot of water. Not only is the volume of the second-biggest watercourse in the country falling every day due to over-extraction for agriculture and domestic use, but industrial tipping is dramatically worsening its plight. In the Mississippi River, one of the longest in the world, marine life has been dramatically reduced due to several previous oil spillages and high levels of nitrogen-based fertiliser runoff, which reduces the oxygen levels in the waters. The constant release of waste into the water has created alarmingly high levels of benzene, mercury and arsenic. The less well-known Passaic River, which flows for 80 miles before discharging into Newark Bay, pays a daily price for decades of industrialisation in the region. Dangerous organisms which have arrived from overflowing sewers have combined with dioxin, mercury and polychlorinated biphenyls (PCBs) to produce a truly horrific pattern of river sediments.

East Asia has more than its share of candidates for the toxic top ten. The Citarum River in Indonesia, with its world-leading levels of mercury in the water, is polluted for its entire length between its source and its outlets, and its heavy pollution is thought to account for around 50,000 excess deaths annually. In China, the Yellow River has been so polluted by industrial discharge that the water is now even too toxic for agriculture. A worrying rise in the rates of cancer and birth defects has now caused authorities to try and dissuade people from drinking the river water at all.

In South America, the River Plate, the continent's second-biggest watercourse, which flows into the estuary between Argentina and Uruguay, has become highly polluted due to the sluicing of the water with chemical waste from farming activities, and the

Matanza-Riachuelo River, also in Argentina, is regarded as potentially the most polluted river in South America. The slaughter houses and tanneries along its banks give it its nickname of the Slaughterhouse River as well as an extremely unpleasant smell. As if this were not bad enough, the high level of heavy metals in the river, including mercury and lead, are contributing to high levels of cancer among the human population.

Sadly, the magnificent River Nile in Egypt has also made it onto the list of shame. The fact that the most common diseases in Egypt are estimated to be bacterial diarrhoea, schistosomiasis and typhoid fever gives us a feel, from previous chapters, about the river's level of contamination. While most sewage release takes place in Lower Egypt, pollutants are derived from sources along its length, including industrial wastewater, oil pollution and agricultural drainage. The problems are intensified by over-extraction from the river for irrigation and the fact that saltwater incursion in coastal areas is worsening because of rising sea levels due to global warming. All this is added to the problems that Egypt suffers due to the river's geopolitical importance, as we have already discussed.

Finally, we could not leave out Australia on the other side of the world where the health of the Murray–Darling River is an increasing cause of anxiety. Like the Nile, there are concerns about maintaining sufficient water levels in the main river channel to prevent saline inflows coming upstream. Contaminated stormwater runoff from developed land areas and septic leakage from adjacent housing have resulted in huge damage to the river's native species, with nine now in danger of extinction. At the same time, this imbalance in biodiversity has seen a huge jump in the number of invasive species, showing that it is not just the health of the human population that is at risk from our global pollution.

POISONED LAKES

As with our rivers, lakes across the world are subject to the scourge of human-induced pollution. The undisputed winner of the 'world's most poisoned lake title' goes to Lake Karachay in Russia. This small lake in eastern Russia is located within the Mayak Production Association, a nuclear weapons laboratory and fuel reprocessing plant that was kept entirely secret by the Russian government until 1990. In the 1960s, following a severe drought, radioactive dust from the lake's bed was carried by the wind and irradiated half a million people. By the time Mayak's existence was acknowledged, the surrounding area of Chelyabinsk had seen a 20 per cent increase in incidences of cancer, a 41 per cent increase in leukaemia and a 25 per cent increase in birth defects. The lake is now full of concrete to keep radioactive sediment away from the shore, where the river-banks will be dangerous for centuries to come.

Lake Victoria is Africa's largest lake, the largest tropical lake in the world, and the world's second-largest freshwater lake in terms of surface area. A renowned tourist destination because of its spectac-ular beauty, plush hotels and lush tropical islands, the lake has over 200 fish species, which makes it the most sought-after destination for fishing in Africa. Yet, this African jewel is under spectacular attack from pollution. The populations of Uganda, Kenya and Tan-zania have almost no access to sewage infrastructure, and so a per-petual flow of sewage, fertilisers and domestic and industrial waste is slowly destroying what has long been the main provider of food and livelihoods in this part of east Africa. Tonnes of plastic bottles and invasive aquatic weeds from Rwanda, Burundi and Tanzania are discharged into the lake by the River Kagera, 400 km from its source in Burundi. A revolting historical note is that during the 1994 genocide in Rwanda, the bodies of those who were massacred were carried into the lake, becoming a serious health hazard for the population of Uganda.

The Great Lakes in North America hold about 90 per cent of the US's freshwater, provide drinking water to over 40 million people in the US and Canada and are home to 3,500 species of plants and animals. Increasingly used as a waste disposal site, the Great Lakes basin became heavily polluted by the late 1900s. As with many similar systems around the world, toxic chemicals, pesticides and heavy metals were dumped into the water by the increasing number of industrial sites around the lakes and are combined with agricultural runoff and the ever-increasing scourge of microplastics. Some pollutants can persist in the ecological food chain for many years, unable to be broken down by biological processes. Phosphorus or PCB, for example, can accumulate in fatty tissue, increasing their concentrations as they move through the food chain. A recent study from the University of Toronto has found that 90 per cent of water samples taken from the Great Lakes in the past decade contain microplastic levels that are unsafe for wildlife. Lake Erie has become the most polluted of all the Great Lakes. Dead fish began to appear along its shoreline in the 1960s and it became yet another victim of eutrophication as massive nutrient enrichment, largely from agricultural runoff, resulted in huge algae blooms, foul taste and odour problems, and fish die-offs due to the lack of oxygen in the water.

Less well known to the outside world is Lake Onondaga in central New York, adjacent to the city of Syracuse. It is a classic tale of pollution worsening over more than a century, with profound consequences for humans and wildlife until determined political and environmental action began to change the prognosis. As industrialisation in America gathered pace, the abundance of salt brine and limestone in the lake's watershed area (used to produce glass, chemicals, detergent and paper) saw rapid economic development. The widespread dumping of industrial and domestic waste led to the banning of ice harvesting as early as 1901, the banning of swimming by 1940 and the banning of fishing by 1970, largely because of mercury poisoning. In 2017, a fifteen-year multi-stage programme to improve the quality of the lake was completed and today, more than sixty-four species of fish now swim in Onondaga, compared to only around a dozen at the lowest point of the lake's history. Change is possible when the will is there.

POLLUTION SANS FRONTIÈRES

All the river pollution that we have previously outlined, including the fertilisers, pesticides and heavy metals that come from agricultural and industrial production ends up in the seas and oceans. Add to this the marine debris, scattered human rubbish floating on, or suspended in, the ocean and which often washes up on our seashores as 'beach litter'. Add to this the mountain of plastics and microplastics that we discard and the pollution that comes from the worldwide shipping industry. On top of this, add the air pollution which carries particles of iron, nitrogen, sulphur, pesticides or dust into our surrounding waters and add again the toxic chemicals which end up in the food chain, leaving us able to poison many of our planet species, including humans, and we can see just what a mess, a disgusting mess, we are making of the one resource that makes our planet unique.

Around 80 per cent of the waste in the seas and oceans comes from land-based activity, with the seas around the continental shelves being particularly vulnerable. Although ocean pollution is heaviest near the coasts and is most highly concentrated along the coastlines of low- and middle-income countries, pollution can also be found on remote islands and even the deepest ocean trenches, truly 'pollution sans frontières'.

Coastal pollution, usually because of rivers loaded with agricultural runoff, sewage and industrial waste can produce harmful algae blooms that create red, green or brown tides. The blooms themselves can cause powerful toxins that accumulate in fish and shellfish and they can be extremely harmful to humans. Likewise, mercury can increase to higher levels in specific types of fish, such as swordfish or shark. While this can be especially dangerous for expectant mothers, exposure to mercury can cause a range of damaging effects on human health. One study among 1,800 men found that those with the highest levels of mercury were twice as likely to die from heart-related problems than men with lower levels, while other studies have linked it to increased risks of Alzheimer's, Parkinson's and other neurological and psychological diseases, such as depression and anxiety.

The major culprit of producing dangerous levels of mercury in the seas and oceans is the burning of coal in domestic homes and industry. As the mercury vaporises, it enters the atmosphere and, since what goes up must come down, it is eventually washed back into the sea. While seawater itself contains only small concentrations of mercury, due to a process called biomagnification, sea plants like algae absorb the mercury, fish eat the algae and predatory fish then eat the smaller fish. As a result, the larger, predatory fish can have mercury concentrations up to ten times higher than the fish they consume. This is why it is advisable not to eat swordfish or shark more than once a week.

A major problem is the length of time that these pollutants remain in the oceans. Persistent toxins, those which do not biodegrade quickly in the marine environment, not only include heavy metals like mercury, lead and arsenic, but also pesticides, phenols, dioxins, PCBs (from paints, glues and plastics) and insecticides such as dichloro-diphenyl-trichloroethane (DDT).

Other forms of pollution, such as oil spills, can cause long-term injury to the marine environment and seriously damage the process by which microorganisms convert atmospheric CO_2 into oxygen, thus damaging the ability to remove greenhouse gases from the environment. Another issue about which we should be concerned is the growing evidence that our polluted waters may be a suitable breeding ground for microorganisms, whose numbers may pose an increasing risk to human health if swallowed directly or incorporated into the food chain.

And then, of course, there are plastics.

THE CURSE OF THE PLASTIC AGE

The Great Pacific Garbage Patch, also known as the Pacific trash vortex, spans waters from the west coast of North America to Japan. The patch is actually composed of two entities: the Western Garbage Patch, located near Japan, and the Eastern Garbage Patch, located between the US states of Hawaii and California. Reading about and seeing this revolting phenomenon was one of the things that motivated me to write this book.

The facts on plastic pollution are truly shocking. At the current rate of accumulation, plastic is expected to outweigh all the fish in the sea by 2050. In the first decade of the twenty-first century, we made more plastic than had been made in the whole of history up to that point, and there are now somewhere between 15 and 50 trillion pieces of that plastic in the world's oceans. There is now nowhere on the surface of our oceans, or anywhere on the planet, that is plastic-free. Microplastics have even been positively identified in Antarctic snow.

The entire Great Pacific Garbage Patch is bounded by the North Pacific Subtropical Gyre (NPSG). The National Oceanic and Atmospheric Administration (NOAA) defines a gyre as a large system of swirling ocean currents in which the circular motion of the gyre draws debris into this stable centre where it becomes trapped.

These patches are almost entirely made up of tiny bits of plastic, called microplastics and this 'soup' is intermixed with larger items, like discarded fishing gear and old shoes. In fact, a 2018 study found that synthetic fishing nets made up nearly half the mass of the Great Pacific Garbage Patch.

The seafloor beneath the patch is also in danger of becoming an underwater trash can as oceanographers and ecologists recently discovered that about 70 per cent of marine debris sinks to the bottom of the ocean.

Plastic pollution, which makes up around 80 per cent of marine debris, ranges in scale from discarded bottles and bags down to the microplastics, which are created by the gradual photodegradation of plastic waste. The multiple chemicals that are added to plastic to make it colourful, flexible or flame resistant include carcinogens, neurotoxins and the kind of endocrine disruptors that we discussed in relation to rivers.

The world's top emitters of plastic pollution in our oceans are China, Indonesia, the Philippines, Vietnam, Sri Lanka, Thailand, Egypt, Malaysia and Nigeria. It is pretty horrifying to learn that, in fact, China, Indonesia, the Philippines, Thailand and Vietnam dump more plastic into the seas than all the other countries in the world combined. Of these, China is by far the worst culprit. Of the 60 million tonnes of plastic waste produced in China every year, only around 16 million tonnes are recycled, with the Yangtze River pouring more plastic into the world's oceans than any other.

Our precious marine life is hugely threatened by plastic, which can entangle it, suffocate it or block its digestive tracts. Seabirds and turtles have been found with bottle tops in their stomachs and countless numbers have died from respiratory obstruction or plastic ingestion. The discarded fishing nets, ghost nets, which make up such a high proportion of the Pacific Garbage Patch are present throughout other parts of the global maritime environment where they can entangle fish, sharks, dolphins, turtles and seabirds. Once caught, they can starve, become cut and infected or, if they are unable to move, suffocated. Fish in the North Pacific are estimated to ingest between 12,000 and 24,000 tonnes of plastic each year and studies suggest that a quarter of all the fish sold at California markets contain plastic in their guts and it is thought that around 60 per cent of all seabird species have eaten pieces of plastic which accumulate in their stomachs, reducing storage volume and precipitating starvation. It is also estimated that this horrendous figure of 60 per cent will rise to 99 per cent by 2050, such is the rapid increase of plastic in our oceans. Some species, such as Hawaiian monk seals and Steller sea lions, are becoming critically endangered because of the buildup of plastic in the habitats that serve as pup nurseries.

What then can we do to reverse this horrible mess and prevent it from worsening in the future? UN Sustainable Development Goal 14 contains a target to 'by 2025, prevent and significantly reduce marine pollution of all kinds, in particular from land-based activities, including marine debris and nutrient pollution'. We have been successful in cleaning up estuaries and harbours around the world and have restored some coral reefs by determined and consistent action, so we know that change is possible. Since many microplastics are the same size as small sea animals, the option of simply trying to hoover them up out of the ocean is not practical, even if it was affordable (which all estimates suggest it is not). The first targeted action must be for humans to pollute less and to create greater awareness of the dangers of plastic pollution and encourage recycling through positive and negative incentives. The second but probably less effective method is to use technological advances in chemistry to create plastics which are more likely to be biodegradable. Whichever mechanism we follow, we must accelerate our activities and strengthen our commitment before the wildlife in our oceans, with all the consequences for humans as part of the global ecosystem, is irreversibly destroyed.

Of course, not all debris at sea is a result of carelessness or reckless disregard for the environment. In 2009, Francesco Forti, a former member of the Calabrian mafia, known as 'Ndrangheta, confessed to an Italian judge that the criminal organisation had made huge amounts of money from sinking ships carrying nuclear and toxic waste in the Mediterranean in the '80s and '90s. Environmental groups suspect that between fourteen and 100 ships were sunk at the deepest points of the Mediterranean, without ever launching a Mayday signal and without their crews ever being found. According to Forti, Italy's state energy research agency, the Nuclear Energy Agency (NEA), paid the mafia to get rid of 600 drums of toxic and radioactive waste from Italy, Germany, France, Switzerland and the US, with Somalia as the destination. Accusations were made that other developing countries were also destinations for toxic cargoes that either 'disappeared' or were buried on land with the acquiescence of local politicians. How reliable Forti's testimony was, and whether authorities will ever have the will to follow his accusations to their logical conclusion, are questions that may never be answered.

SAD SEAS AND OCEANS

In terms of our named bodies of water, which are the most polluted overall? Perhaps unsurprisingly, given that it is the biggest expanse of water on the planet, the Pacific Ocean is regarded as the most polluted. On the whole, as well as having the largest patch of plastic debris, the Pacific is thought to have the largest amount of sewage pouring into it and from 2003 to 2012, the number of chemical pollutants that entered the Pacific Ocean increased by 50 per cent.

The Indian Ocean is regarded as the second-most polluted, with 1.3 trillion plastic pieces, mostly within its own Indian Ocean garbage patch. This is in no small part due to the fact that two of the world's most polluting rivers, the Ganges and the Indus, flow directly into it. As well as its plastic problem, around 40 per cent of all petroleum spills are thought to happen here, including the recent 2020 spill in Mauritius, which saw around 1,000 tonnes of oil leak into the Indian Ocean, with thousands of species left drowning in oil, a nightmare for Mauritius's tourism industry, the economy, food security and health, as well as being an environmental disaster.

The Atlantic Ocean, the world's second-largest ocean, also has horrendous pockets of pollution, including the Gulf of Mexico, which is mentioned separately. The North Atlantic garbage patch, first discovered in 1972, continues to grow and is now believed to be several hundreds of miles wide, with over 200,000 pieces of waste per square mile.

The Gulf of Mexico, surrounded by the USA, Mexico and Cuba is one of the largest dead zones in the world. Gigantic amounts of nutrients, especially phosphorus and nitrogen, flow down the Mississippi River from farms, lawns and industry along its route into the Gulf. The algae blooms, which we have already considered in relation to rivers and lakes, have created hypoxic water where fish find it difficult, or even impossible, to survive. The National Oceanic and Atmospheric Administration (NOAA) estimates that the dead zone costs the US tourism and seafood industries over $80 million a year, with, for example, fishermen having to travel further and further away from the dead zone to make their catches.

The Baltic Sea is one of the most polluted seas in the world. Its catchment area is about four times the size of the sea itself, corresponding to about half the size of Europe. Rivers carry large amounts of nutrients and hazardous contaminants into the Baltic Sea and, along with oil spills, this has meant more than half of all its fish species being categorised at critical biomass levels, threatening the maintenance of biodiversity. On top of this, substantial airborne contamination dumps heavy metals and hazardous organic substances onto the water's surface.

Perhaps its most frightening pollution, however, lies on the seabed. An almost unbelievable 40,000 tonnes of wartime munitions lie scattered there, including bombs, missiles and chemical agents, which were abandoned there at the end of the war. This was considered a safe and cheap solution when the Allies dumped unwanted weaponry in 1945, but they have slowly been leaking toxic chemicals, including TNT, mustard gas, arsenic and phosgene, into the Baltic ever since. These have destabilised ecosystems with the ever-present fear that the cancer found in many species of marine life may find its way into the human population.

The Mediterranean, so popular with tourists, is also hugely polluted. Plastics account for over 95 per cent of the total floating litter and more than 50 per cent of the litter on the seabed. On its popular beaches, single-use plastics represent over 60 per cent of the recorded marine litter. Fertiliser and pesticide use is above the global average in over half of the Mediterranean countries and 184 million tonnes of solid waste flow into the sea per year. Forty-nine per cent of coastal water bodies in the Mediterranean have failed to achieve good environmental status, with heavy metal pollution combining with both treated and untreated wastewater, industrial effluents and the runoffs from intensive farming and aquaculture.

Even the Caribbean Sea, the poster child for sandy beaches, palm trees and beautiful sunsets, is now one of the areas worst affected by human activity, with oil spills, waste and chemical pollution and overfishing all contributing to a disturbing reduction in the marine life which has been one of the chief attractions to global tourists for so long.

Wherever we look, then, the picture is the same. If we cannot reduce the human population, then we need to improve its behaviour, requiring individuals to take greater responsibility for their own actions and cutting overall pollution. Technology and innovation can only do so much to correct or mitigate the impact of human behaviour.

Of all the environmental challenges we face, the health of the oceans is right at the top of the list. A quick look at what we have already done to our marine environment should disgust sufficiently enough to want to change our behaviour, and quickly.

CHAPTER 21

WATER AND CLIMATE CHANGE

SALINITY AND DESTINY

Salt has two important effects on water. The first is that it makes seawater denser than freshwater. The second is that saltwater needs to be colder than freshwater before it freezes. Salinity is measured in parts per thousand (ppt) with freshwater having a salinity value of less than 0.5 ppt while salt water has an average salinity of 35 ppt. Water with a salinity of 17 ppt freezes at around -1°C (30°F) while water at 35 ppt freezes at about -2°C (28.5°F).

As we will see, both of these effects are important in the generation of ocean currents.

A salinity of 35 ppt means that about 3.5 per cent of the weight of seawater comes from the dissolved salts. In a cubic mile of seawater, the weight of the salt (as sodium chloride) would be about 120 million tons.

While this is an average concentration, as we discovered in Part 1, there are large variations in salinity throughout the oceans. Where there are strong winds that aid evaporation but little rainfall, the oceans tend to have a higher salinity, such as in the North and South Atlantic. The Mediterranean is even higher, with around 38 ppt or more, a result of being almost completely closed off from the main ocean with only the Strait of Gibraltar connecting them.

Conversely, there are parts of the oceans where salinity is much lower. The tropics, being near the equator, receive the most consistent rainfall and this falling freshwater decreases the salinity of the surface water there. Around the Arctic, salinity is down to 30 ppt in some places, as icebergs that have broken off ice sheets formed on land do not contain salt and so the freshwater that is released as they melt reduces the salinity of the ocean around them. The Baltic Sea, which is almost completely enclosed by northern Europe and Scandinavia, has a very low salinity of around 10 ppt. Ice annually occurs in the Baltic Sea for seven months, from November to June, due to its low salt concentration, which occurs because of the high freshwater runoff from the hundreds of rivers flowing into it. This allows the water to freeze at a higher temperature, thus creating more ice that lasts longer.

It is, incidentally, easier to swim in water with a higher salinity. The Mediterranean, for example, is much easier to swim in than the Baltic, and not just because it is a good deal warmer. The weight (and downward pull) of a swimmer is balanced by the less immersed parts of the body, an effect which is greater where the water is denser. In the Dead Sea, with a salt ppt of an incredible 280, it is possible to float in the water and read a newspaper. I know, because I have tried it!

THE GLOBAL CONVEYOR BELT

The global ocean has an interconnected circulation system which is powered by wind, tides, the earth's rotation, solar energy from the sun and differences in water density.

While winds drive the currents in the upper 100 m or so of the ocean's surface, it is the differences in water density that drive deeper currents.

The density of seawater is influenced by both its temperature (thermo) and its salinity (haline), with the former having the greater effect. Density increases as salinity increases and temperature decreases. Thus, the great driver of global currents is known as the thermohaline circulation (THC).

Also known as the global conveyor belt, the thermohaline circulation is much slower than the more superficial, wind-driven currents, with a typical speed of 1 centimetre (0.4 inch) per second. The biggest difference, however, is that this slower flow extends to the seafloor and forms circulation patterns that envelop the global ocean.

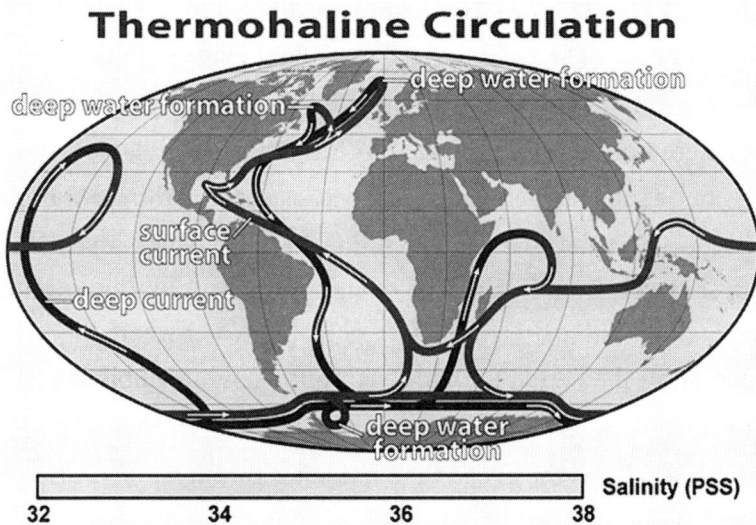

Thermohaline Circulation

deep water formation

deep water formation

surface current

deep current

deep water formation

Salinity (PSS)

32   34   36   38

The concept of the thermohaline circulation is a relatively recent one, with the theory being described in 1960 by the American scientists Henry Stommel and Arnold Arons. The system works by combining two simultaneous processes, with warm surface currents carrying less dense water away from the equator towards the poles while cold deep ocean currents carry denser waters away from the poles towards the equator. During the respective northern and southern winter seasons, cooling or net evaporation causes surface water (with a lower temperature and higher salinity) to become dense enough to sink. Surface water, in turn, is pulled in to replace the sinking water and it too eventually becomes cold and salty enough to sink. The basic thermohaline circulation is, therefore, one of the sinking of cold water in the polar regions, chiefly in the North Atlantic and near Antarctica. These dense water masses spread into the full extent of the ocean and gradually upwell again to feed a slow return flow to the sinking regions. This global circulation system is vital for distributing heat energy, regulating weather and climate and circulating vital nutrients and gases throughout the oceans.

In the case of the oceans, the continual excess of evaporation versus precipitation would eventually leave the oceans empty if they were not being replenished by additional means. Not only is this continually happening, largely because of runoff from the land areas, but over the past 100 years, they have been *over*-replenished with sea levels around the globe rising approximately 17 cm over the course of the twentieth century.

Sea levels have risen both because of the warming of the oceans, causing water to expand and increase in volume, and because more water has entered the ocean than has left it through evaporation or other means, a subject we will explore in more detail shortly.

THE ATMOSPHERE

The amount of water in the atmosphere at any moment in time is only 12,900 km³, which is such a minute fraction of earth's total water supply that even if it were to completely run out, atmospheric moisture would cover the earth's surface to a depth of only 2.5 km³. However, far more water – in fact, some 495,000 km³ of it – are cycled through the atmosphere every year. The effect is the equivalent of removing and replenishing the entire amount of water in the air around forty times a year.

With so much of the climate debate focused on carbon dioxide, many people are unaware that water vapour is the most abundant greenhouse gas in the atmosphere, both by weight and by volume. Not only is water vapour abundant, but it is also an extremely effective greenhouse gas, absorbing longwave radiation, radiating it back to the surface and so contributing to the warming effect.

When compared to other greenhouse gases, however, water vapour remains in the atmosphere for a much shorter time, usually days, before precipitating out. Other greenhouse gases, such as carbon dioxide or methane, can remain in the atmosphere for much longer periods, years or even centuries, and so contribute to the warming effect for a much longer period.

Increased water vapour in the atmosphere is part of a natural feedback process. Warmer air can hold more moisture and as the climate warms and air temperatures rise, there will be a higher rate of evaporation from both land and water sources, causing an increased water content in the atmosphere. Since water vapour is such an effective greenhouse gas, this contributes to further warming and enhances the greenhouse effect.

To put the relative effects into perspective, the water vapour feedback process is estimated to be responsible for a doubling of the greenhouse effect when compared to the addition of carbon dioxide on its own.

How does this natural greenhouse effect interact with that produced by human activity?

Our modern civilisation has been able to develop because this natural greenhouse effect has kept the earth's temperature at an optimal level for human development. It is controlled by non-condensable gases, mainly carbon dioxide, with methane, nitrous oxide and ozone playing a smaller part. From the mid-twentieth century onwards, new substances such as chlorine and fluorine – containing solvents and refrigerants – have been added, but because these are not condensable at atmospheric temperatures and pressures, far greater amounts can be packed into the atmosphere. These, along with the carbon dioxide that has been building up since the Industrial Revolution, and exacerbated by an ever-greater human population, have led us to the predicament we face today. The addition of non-condensable gases is a crucial part of the problem as the temperature rise they produce has led to the increase in water vapour, which has further raised the temperature. This is what is known as a positive feedback mechanism.

As the earth warms, the global water cycle is amplified, with increased evaporation and, ultimately, increased precipitation. It is estimated that the water cycle has accelerated by 2 to 4 per cent per degree Celsius since 1960.

RISING SEA LEVELS

Today, the sea level is 10–20 cm (4–8 inches) higher than it was a century ago, largely because of this advancement of the water cycle. It is expected to rise further during the twenty-first century. It has been estimated that if we are able to minimise the effects of climate change, then the rise in sea levels could be contained to around 30 cm (12 inches). If not, we could see the level rise by up to a metre (36 inches). While there can be no certainty about these predictions, we can extrapolate findings from recent experience and the measurable increases in both temperature and sea levels.

About 90 per cent of the excess heat trapped by atmospheric greenhouse gases is eventually soaked up by the world's oceans, thus producing the higher temperatures that can affect global systems like ocean currents. The sheer size of the oceans means that we can easily underestimate the importance of the relatively small rise in temperature (just over 0.5°C) in the past century. The effects can already be felt, and the warming process is accelerating.

As global temperatures rise, water from melting glaciers and ice sheets flows down rivers and is added to the volume of the ocean. Over the past 100 years we have been able to observe and measure how mountain glaciers, Arctic glaciers and Greenland's ice have all dramatically reduced in size. This melted ice must go somewhere and it ends up in the oceans, producing a rise in sea level. Incidentally, contrary to what some seem to believe, melting ice that is already in the ocean does not produce a rise in sea level.

A second effect is that ocean water expands as it warms, increasing its volume, so that the water in the ocean takes up more space and the sea level is higher. It is thought that since the mid-twentieth century, over 90 per cent of the excess heat held in the atmosphere has made its way into the ocean. Without this natural absorption it is likely that the effects of climate change would have been more dramatic. The downside is that the expanding warming water, and consequent rises in sea level, have produced coastal flooding, threats to sensitive marine life and the potential obliteration of island communities.

Thermal expansion and melting ice have each contributed about half of the recent sea-level rise, though there is some uncertainty about the exact magnitude of the contribution from each source. The thermal expansion of seawater is predicted to account for about 75 per cent of the future rise in the sea level according to earth system models.

How much of the rise in sea level then is accounted for by thermal expansion, and how much is accounted for by melting ice?

It is now estimated that thermal expansion raised sea levels by an average of 1.1 mm per year from 1993–2010, which accounts for much of the overall rise we have seen. This compares to a contribution of 0.86 mm from melting glaciers, 0.3 mm from the Greenland ice sheet and 0.27 mm from the Antarctic ice sheet. So, how much will the sea level rise as the climate warms? The current consensus is that for every 1°C rise in temperature, we will have a rise in sea level of 15–20 cm (6–8 inches). That is the rise in both temperature and sea level that we have seen over the past century, with acceleration beginning in 1993.

Warming global waters also have an influence on the atmosphere above. The increased energy that they contain is associated with more powerful hurricanes and tropical cyclones, a phenomenon that we are witnessing with increased frequency. Less well recognised is the fact that warmer water can dissolve less carbon dioxide, resulting in more remaining in the atmosphere and causing the global warming process to accelerate. The carbon dioxide that has already been absorbed into the oceans has produced an acidification effect. The carbonic acid that forms when carbon dioxide is dissolved in seawater has reduced the potential hydrogen (pH) in the top layer of the oceans from about 8.2 to 8.1. While this doesn't sound particularly alarming, because pH is measured on a logarithmic scale, it represents almost a 30 per cent increase in acidity, a significant shift in seawater chemistry that could have a devastating impact on marine life.

Just as we encounter heatwaves on land, the oceans can have their own heatwaves too, and these can also have huge consequences for marine life. While these tend to only last for a few days, from 2013–2015, a marine heatwave known as 'the blob' is estimated to have killed a million seabirds on the west coast of the United States. Rising water temperatures can also cause coral bleaching, damaging important coral reef ecosystems, and can also cause the mass migration of marine species that are looking for the right conditions in which to feed and spawn. If we add to this the potential effect on the growth of most fish and cephalopods, such as squid and octopuses, the repercussions for humans who depend on fish as their chief source of protein could be devastating.

EFFECT OF CLIMATE CHANGE ON SALINITY

The saltier parts of the ocean have increased in salinity by 4 per cent in the fifty years between 1950 and 2000 and if the water cycle continues to intensify, as it has in recent years, this trend is likely to continue. Does this matter? In short, yes. Saltier oceans produce warmer climates, largely due to the changes they induce in ocean dynamics. It is estimated that a rise in ocean salinity from 20 to 50g/kg would produce a 71 per cent reduction in sea ice cover since salt depresses the freezing point of seawater and may inhibit the formation of sea ice in salty oceans.

Changes in ocean salinity are a sensitive reflection of the net exchange of freshwater between the ocean and the atmosphere and so can be used to estimate changes in the water cycle. In short, the fresh gets fresher and the salty gets saltier in much of the ocean. Any amplification of the water cycle is likely to result in drier regions getting even drier with worsening and more frequent droughts affecting livestock and crops and creating the conditions for increasingly destructive wildfires. At the same time there will be a heightened risk of heavy and extreme rains producing more flooding and more powerful hurricanes and cyclones, which will particularly affect coastal communities already struggling with rising sea levels.

It is not, of course, just humans living on the land who will be affected by these trends, but predicting how much damage can be done to marine ecosystems because of changes in salinity is far from an exact science.

We have a reasonably good understanding of how alterations in temperature, acidification and nutrients will affect ocean and coastal ecosystems, yet even though salt exposure levels are critical for many organisms, our knowledge about the effects of human-induced changes in salinity is patchy.

Earlier, we discussed the impact, in terms of human health, of changing salt concentrations and the impact that they could have at a cellular level. Much the same applies to all living organisms, including fish.

A freshwater fish in seawater will experience high cellular damage caused by overpressure in its cells and a seawater fish in freshwater will become dehydrated since its cells are incapable of retaining water.

In recent decades, evidence points to the huge damage that rapid changes in salinity can cause to marine life. In Spain, massive numbers of shellfish deaths were reported in the estuaries of Galicia due to sharp decreases in salinity resulting from heavy rains, freshwater releases from dams and increased river runoffs.

In freshwater environments, zooplankton and amphibians and

freshwater mussels in early life stages have been found to be highly sensitive to increased salinity in most species of fish with egg fertilisation and incubation, development, and growth thought to be affected by changes in salinity, although it is unclear how much this is dependent on other factors, such as an increase in water temperature.

While the ability to calcify the exoskeleton in crustaceans, such as crabs, lobsters and shrimp or the shell in molluscs, depends on the interaction of several different factors, including salinity and pH, again, the exact science is poorly understood.

Vital plants can also be affected. Seagrasses, which are essential in the production of photosynthesis, can also be massively adversely impacted by rises in salt levels.

A lot of research needs to be undertaken in this key area, but the risks are becoming increasingly clear.

EFFECT OF CLIMATE CHANGE ON CURRENTS

As we have discussed, the thermohaline circulation carries oxygen from the surface to the deep ocean and, similarly, brings nutrients from the deep to the surface. It is also important in bringing heat to higher latitudes and polar regions. The Atlantic Ocean conveyor belt, or the Atlantic meridional overturning circulation (AMOC) heats northern Europe as it extends to the Greenland and Norwegian Seas, pushing back the winter margin of sea ice. It is this phenomenon that prevents Britain from enduring freezing continental winters, as the AMOC raises the temperature by about 5°C.

The system, which includes the Gulf Stream, carries heat north from the tropics and southern hemisphere until it loses it in the northern North Atlantic and the Norwegian and Labrador Seas, which in turn leads to the deep sinking of the colder waters.

The AMOC influences weather in Europe and eastern North America, and it is thought that a weakened AMOC would be associated with increased winter storms and summer heatwaves in Europe and extremely cold winters and intense blizzards in eastern North America.

Does climate change affect the great global conveyor belt and, if so, by how much and with what implications?

In 2015, scientists determined that the potential collapse of the cod fishery in the Gulf of Maine was due to rapidly warming water which fitted the pattern of a slowing Atlantic circulation.

It is clear that there is considerable variability in the AMOC, both seasonally and from year to year, but there is also strong evidence that it is getting weaker. Sea surface temperature analysis, dating back to 1901, confirms the trend and sediment studies from the bottom of the Labrador Sea suggest that the currents bringing cold water back south have been slowing since 1750. Recent evidence also indicates that the overturning thermohaline circulation near Antarctica is also slowing. As increasing amounts of ice melt, the surface waters become less salty and less dense and so are less able to sink to greater depths. Some studies also imply that if global warming continues at its current rate, then the Antarctic overturning will slow down by more than 40 per cent in the next thirty years. This means that nutrients will become trapped in the deep ocean rather than being brought back to the surface waters, creating a catastrophic impact on marine ecosystems.

In the North Atlantic, the AMOC is predicted to slow by 20 per cent in the next few decades judging by current trends. Since it currently keeps Europe mild, there would be a considerable reduction in temperature. Paradoxically, global warming would cause significant cooling in this part of the world, although it is fair to say that there are considerable differences among experts over how quickly change may occur and what its extent may be.

What is clear is that more freshwater is definitely entering the system.

In 2016 the Arctic lost about 23,600 square miles of ice daily, compared to the long-term average loss of 18,000 square miles per day. Although this trend seems to have stabilised a little in recent years, the Arctic is warming up twice as fast as the rest of the world. The snows of Siberia have also been melting sooner; more freshwater has come from Greenland and more still has come from north Canada, where ice and snow have been melting at a growing rate.

We know from history that there were huge changes in the AMOC long before humans were producing greenhouse gases. We know that it became weaker in the run-up to the planet's last major cold snap around 13,000 years ago, with a gradual slowdown occurring in the preceding 400 years. There was a similar pattern at the end of the Ice Age with the current strengthening around 400 years before the atmosphere began to heat up. Then, the warming of Greenland by about 8°C in just a few decades saw a mass melting of glaciers and the volume of sea ice diminishing rapidly.

There is no clear consensus about what the current changes mean for the future.

Climate records preserved in Greenlandic ice show how inconsistent changes can be. The last Ice Age, which comprised around twenty-five periods of sudden climatic changes with large temperature spikes over several decades, was followed by gradual cooling in the North Atlantic. Just to make the picture even more confusing, there were also extremely cold periods between some of these warming events. There is some suggestion that rapidly melting ice injected huge amounts of freshwater into the North Atlantic, slowing the AMOC and causing considerable cooling in the regional climate. It is almost impossible to know in any detail how a major slowdown in the AMOC would affect our weather patterns today. We cannot say whether this would outweigh the consequences of overall global warming or where the balance would lie.

There is also the question of how, with our already diminishing ice sheets, a large enough volume of ice could be injected into the sea to produce the conditions needed for sudden cooling. Still, it is a reminder about how little we actually know about, rather than trying to predict through modelling, the complex interactions that govern the climate on our planet.

CHAPTER 22

DESALINATION

Uniting several of these issues brings us to the topic of desalination where, to boost the availability of freshwater, salt is removed from seawater, with other minerals being removed in the process. It is, in a world with a growing demand for freshwater, one of the few rainfall-independent resources available. It has also become a more attractive option as the costs of the technology have fallen.

Advancements in reverse osmosis technology have brought desalination costs closer to other alternatives. Ten years ago, desalinated water cost more than $9 per 1,000 gallons, but today, the range is $2 to $5 per 1,000 gallons.

More than 20,000 desalination plants are now in operation worldwide, producing a combined total of more than 100 million m³ of drinking water per day.

Most Gulf countries now largely depend on desalinated water for their inhabitants' consumption, making it the region most heavily invested in the technology. In the United Arab Emirates (UAE), 42 per cent of drinking water comes from desalination plants producing more than 7 million m³ per day; in Kuwait it is 90 per cent; in Oman 86 per cent and in Saudi Arabia 70 per cent. For any country thinking about the widespread adoption of desalination, a closer look at detailed costs might provide some uncomfortable food for thought. One of the biggest costs is not the desalination process itself, but transporting the water to its ultimate destination. It is estimated, for example, that to transport desalinated water to a high altitude, such as to Mexico City or long distances inland, such as to New Delhi would be equal in cost to the desalination process itself, making it prohibitively expensive to introduce.

There are several ways in which desalination can take place.

Distillation, the traditional process of desalination, involves boiling and recondensing seawater to leave salt and impurities behind.

A new approach, reverse osmosis, operates by pushing saltwater, under high pressure, through a semi-permeable membrane whose pores are too small for the salt molecules to pass through. As the membranes are both expensive and fragile, various pre-treatment stages are used to prevent damaging them, thus removing contamination by calcium and magnesium, insoluble particulates, bio filing and scaling, as well as bacteria and viruses. Yet, the purity of the desalinated water can also have its disadvantages. The removal of substances like magnesium can have deleterious effects on humans, as we discussed in the chapter relating to Israel's desalination programme, to the extent that magnesium is having to be reintroduced in the form of supplements to correct the imbalance.

Although reverse osmosis is now the increasingly preferred technology, with hindsight, its advance was never as certain as it might appear. From the point when its advantages were scientifically and practically established, it still took reverse osmosis about three decades to overtake thermal desalination as the market-leading technology.

Apart from the energy consumption and costs involved in the production of desalinated water, there is now increasing concern about the impact of toxic brine, the byproduct of the process. In most desalinisation processes, for every litre of potable water that is produced, around 1.5 L of liquid, polluted with chlorine and copper, is created.

Globally, desalination plants are now estimated to discharge 142 million m³ of hypersaline brine every day, a 50 per cent increase on previous assessments. It has been calculated that this is sufficient to cover Florida under 30.5 cm each year.

Moreover, the geographical concentration is alarming. Fifty-five per cent of global brine is produced in just four countries: Saudi Arabia (22 per cent), UAE (20.2 per cent), Kuwait (6.6 per cent) and Qatar (5.8 per cent). Middle Eastern plants, which largely operate using seawater and thermal desalination technologies, typically produce four times as much brine per cubic metre of clean water as plants where river water membrane processes dominate, such as in the US.

Desalination plants near the ocean (almost 80 per cent of brine is produced within 10 km of the coastline) most often discharge untreated waste brine directly back into the marine environment. Because it is heavier than seawater, the brine tends to settle towards the bottom of the coastal areas where it is released – unless it is diluted. The excess salt decreases the amount of oxygen dissolved in the water, thus suffocating animals on the seafloor. Technologies exist to reduce brine waste prior to disposal or to mine pollutants out of the waste for commercial use – but this is typically cost-pro-hibitive. Instead, plants use other strategies to minimise damage.

One such alternative involves situating plants in areas where strong currents help to disperse the brine, but this is not always possible. For example, the Arabian Gulf is shallow, lacks strong currents and has seen incoming freshwater slow to a trickle due to upstream dams and people in the region diverting water for drinking and irrigation. The Gulf is also a receptacle for salty 'produced water' from the oil and gas industry. As a result of these factors the Gulf is now about 25 per cent saltier than typical seawater, with hotspots double or triple its regular salinity.

Some plants make efforts to mix the brine more effectively into the ocean when it is discharged, either by using multiple waste outlets that spread it over a larger area or by pressurising the waste flow to disperse it by force. In Australia, where the Millennium Drought, the country's worst in living memory, created concerns that the country could run out of freshwater, six major desalination plants were constructed, the largest being in Victoria state. The high capacity at which the plants were run prompted concerns about the disposal of brine and the impact upon coastal ecosystems. One of the solutions was to release brine at high pressure in an attempt to spread it further and diminish its environmental impacts. A recent study, conducted over six years at the Sydney Desalination Plant in Australia, found its pressure diffuser reduced excess salinity in coastal areas where waste brine was released, but the abnormally fast flow prevented species with slow-swimming larvae – such as tube worms, lace corals and sponges – from colonising the impact zone. At the same time, species that thrive in high-flow conditions – such as barnacles and bivalves – increased in number.

The question is whether brine could actually be used in a way that turns it into a useful resource rather than an environmental poison. Recent studies have, optimistically, shown that the brine waste can be converted into useful chemicals, including those that can make the desalination process itself cleaner and more efficient. The production of sodium hydroxide can pre-treat seawater going into the plant, changing its acidity and preventing the fouling of the reverse osmosis membranes, a major cause of failure and increased cost. Hydrochloric acid can also be produced which can, again, be used for cleaning parts of the desalination plant but is also widely used in chemical production and, potentially, as a source of hydrogen.

As is so often the case, the technology that produces solutions often creates new questions of its own but, with a booming global population and a limited supply of freshwater, it will be our ingenuity and technological prowess that will be needed to create a safe pathway to our future. Or else!

AND SO... TO THE MOON?

Although it is only the fifth largest of over 200 moons in our solar system, the moon is our planet's only natural satellite, a perpetual celestial partner, constantly orbiting earth in an elliptical path.

In 1645, the Dutch astronomer Michael van Langren published the first known map of the moon and its 'maria', the Latin term for sea. This documented, for the first time, the widely held contemporary view that the dark spots on the lunar surface were oceans. Although we now know that these are in fact basalt plains created by early volcanic eruptions, around 4 million years ago the moon could have had enough atmosphere and liquid water on its surface to have real seas and pressurised regions in the moon's interior might still contain liquid water. While the earth is tilted 23 degrees away from the plane in which planets orbit the sun, the moon has much less of a tilt. This means that while we have seasons, at the lunar poles, large craters with high rims block sunlight from ever reaching the interior, as they have done for billions of years. These areas are called permanently shadowed regions (PSRs) and it is postulated that they could contain large amounts of liquid ice. Their presence could revolutionise our view on the habitability of the moon as it could be mined for liquid water and for potential horticulture. It might also be used to provide the basis for rocket fuel, enabling further planetary exploration.

Our knowledge of water on the moon has improved enormously in recent years. The major, definitive, breakthrough came in 2008 when India's Chandrayaan-1 spacecraft carrying NASA's Moon Mineralogy Mapper (M3) determined that there were ice molecules inside the moon's polar craters. In 2009, NASA deliberately smashed part of its Lunar Reconnaissance Orbiter (LRO) into the moon's south pole and detected 155 kg of water in the debris plume. More recently, in 2018, data from the Moon Mineralogy Mapper confirmed the location of water ice in some of the PSRs. Revised analysis of moon rock brought back by the Apollo missions has also found hydrogen inside tiny beads of volcanic glass which shows that water existed on the moon when the volcanoes erupted millions of years ago. By extension, if this erupted from volcanoes, it must have come from within the moon, suggesting that water may have been there since its early existence. All of this brings us back to the question of how water would have arrived here in the first place. Much like our own story, water-bearing comets and asteroids would have bombarded the moon, depositing water there. Some of its water may have also come from the earth if the moon was created following a large cosmic collision involving our own planet. While most of the moon's water would have evaporated into space in the absence of an atmosphere, some seems to have found its way into PSRs, where it has been silently preserved for billions of years.

Exactly how much water might exist on the moon has been a subject of intense scientific debate. Observations by the Chandrayaan-1 and LRO instruments suggest that the lunar poles may have over 600 billion kg of water ice while other estimates are both higher and lower. Obtaining precise measurements from landing craft will be a challenge because the steep and rocky terrain of the lunar poles (unlike the flatter areas where craft have already landed) will make predictable and successful landings more physically difficult. Additionally, communication will be more difficult, with any vehicles venturing out of the earth's line of sight and the lack of sunlight meaning that solar panels cannot be used for power with heavier, more powerful batteries, probably relying on some form of nuclear energy being used instead.

Finally, just as there are disputes over the ownership of water on earth, there are likely to be similar disputes on the moon. The United Nations Outer Space Treaty (UNOOSA), while it does not prevent the exploration of lunar resources, prevents the appropriation of them by individual nations with implied restrictions on the ownership of resources. Despite this tentative agreement, some countries, including the United States and Luxembourg, have given their citizens permission to mine and own space resources, setting the scene for what is likely to be a battle over lunar territory, mineral and, most importantly, water rights in the future. It seems that unless we anticipate these issues and act in a concerted way, we are likely to repeat all the errors that we have made on our own planet on that of our next-door neighbour.

CHAPTER 23

WHAT NEXT?

Following Russia's aggressive invasion of Ukraine, the world has been gripped by concerns over energy security and supply. This is despite the fact that for over two decades, there have been numerous warnings about the impending crisis. As ever, global leaders have been too willing to substitute wishful thinking for critical analysis, ignoring the fact that, as history reminds us, hope is a poor basis for policymaking.

How will we deal with the increasing challenge over access to water, one that is more likely to shape the events of the twenty-first century than the competition for oil or gas? If people and nations will fight for fossil fuels, what will they do to control what is, literally, the most vital natural resource of all?

Our supply of freshwater is finite. The water cycle means that water cannot be created or destroyed, but only 3 per cent of the water in our world is the freshwater upon which we all depend. Not only has the human population mushroomed over the past century but it is, increasingly, unevenly distributed, with some of the highest concentrations occurring in areas with the greatest water stress, such as China and south Asia. To add to the problem, climate change is affecting the delicate balance of supply and demand. Today's areas of flood and drought may be reversed tomorrow as glacial melt first increases and then decreases. Today's flooded will become tomorrow's thirsty.

Whether we can navigate these issues successfully or not depends upon the choices that we make. Recent history suggests that we continue to make bad decisions around many of the subjects that we have considered.

The Pacific Institute classification gives three categories of conflict based on the role that water plays in them: where water is the trigger, usually as a result of a dispute over control of the supply, where water is used as a weapon in a violent conflict and where water is the casualty of intentional or incidental violent actions.

Gaza, in a region with long-standing disputes over the use of water from the Jordan River basin, finds itself in a war following the Hamas terrorist attacks on Israeli citizens and Israel's military response. In normal times, Gaza depends on Israel for one-third of all available drinking water, with the rest coming from an increasingly polluted subterranean aquifer and several desalination plants. When the electricity and fuel disappeared, the desalination plants and the wastewater treatment stations shut down. Water was a casualty and, arguably, a weapon in the conflict.

In the related dispute in the Red Sea, Houthi rebels have targeted the vulnerable Bab el-Mandeb Strait, only 16 miles wide at its narrowest point and a key part of the global maritime supply route. The Houthis, who declare themselves to be part of the Iranian-led 'axis of resistance' against Israel, the US and the wider west (along with Hamas and Hezbollah) have already attacked vessels belonging to a number of different nations.

Led by the United States and the UK and joined by the Netherlands, Bahrain, Canada and Australia, the principle of freedom of navigation is rightly being defended by an international coalition.

When this concept became part of the broader body of laws of the sea in 1982, in the United Nations Convention on the Law of the Sea (UNCLOS), Article 87 clearly stated that 'the high seas are open to all states, whether coastal or landlocked', with Article 24 declaring that 'the coastal state shall not hamper the innocent passage of foreign ships through the territorial sea', provided that vessels do not threaten 'peace, good order or security'. The violent and indiscriminate attacks on unarmed shipping, putting human life at risk, is a clear violation of international law.

The economic case for action is also clear. More than 80 per cent of international trade by volume is transported by sea, and around 15 per cent of the total moves through the Red Sea to and from the Suez Canal. About half of this trade is made up of containerised goods, and oil shipments from the Gulf to Europe and North America make up a considerable proportion of the rest. Disruption results in ships having to sail around the southern tip of Africa, adding some 3,500 nautical miles to their journey and culminates in disruptions of ten to fourteen days. Some of the world's biggest companies, including Maersk and Hapag-Lloyd, have already done so, resulting in the delayed delivery of manufacturing components and goods to European markets. In the first week of January 2024, over 220 fewer vessels transited the Red Sea compared to the previous year, and figures suggest that the rerouting of vessels is still increasing. Higher costs in global shipping have already started. For exporting countries like India, for example, the effect will be sharply felt. For a 20-foot container to Europe, shipping costs have gone up to $2,100–$2,500 from the previous $500–$600. Shipping costs to the east coast of the United States have doubled from $1,600 to $3,100 and further price hikes are expected if the dispute continues.

There are other costs rising too, which will eventually find their way to consumers. The price of Brent Crude climbed to over $93 per barrel in the days following the attacks on Israel and is currently trading at around $80, having fallen back to $73 before the Houthi attacks. Disruption to global shipping with backlogs in ports, diversions causing the reduced availability of ships and rising insurance costs will inevitably have a negative economic impact. The environment will also suffer. Having to sail round the Cape of Good Hope requires ships to burn significantly more fuel, and the industry's carbon footprint, which already contributes nearly 3 per cent of all global greenhouse gas emissions, will rise further.

Again, while water may not have been the primary cause of the crisis, it has certainly been used as a means of pursuing a dispute.

As if to add even greater poignancy, nature has made its own contribution to the global maritime problem. On the other side of the world, in Panama, rainfall has been 30 per cent below average over the past year, which has resulted in water levels in the freshwater lakes that feed the canal locks dropping dramatically. The drought is heightened by the El Niño weather phenomenon and has pushed up both fees and congestion at the canal, resulting in fuel tankers and grain shippers taking longer, and more expensive, routes to avoid the problems. We do not have to look far to see how urgent the problems we have discussed now are.

Recent precedents elsewhere do not suggest happy trends for the future. If upstream nations can defy their downstream neighbours despite military threats, what will happen in areas where a single power controls such huge upstream capabilities that it threatens multiple nations downstream simultaneously? Since this is already the case between China and a range of its geographical neighbours, including India, Pakistan and the countries of the Mekong Delta, we need to grasp the immediacy of the problem.

Much more effort at the level of international diplomacy and in global institutions must be directed towards finding peaceful solutions to these potential disputes before they escalate. China will have been looking at the Ethiopian Grand Renaissance Dam project with particular interest, since it appears that Ethiopia has simply been able to face down the threats from its downstream neighbours, even if these were never as real as the rhetoric implied. Whether the case made by the Egyptians and Sudanese is as detrimental as they claim, there can be no doubt that the Ethiopians have now achieved their strategic objective.

If we are to be able to navigate potential conflicts, we need to have international treaties that will be robust enough to deal with what may be rapidly changing circumstances. Agreements will need to deal with changes not only in water supplies themselves but also in the challenges of changing populations, economic activity, climate and infrastructure while allowing for any positive technological changes that may emerge.

The data-sharing elements of this solution merit a special mention. Data, knowledge and understanding are three different things. If they were not, we would not require different words for them in our lexicon. They must all be addressed equally to produce stable outcomes.

To have a sustainable, peaceful and co-operative future, any treaties will have to have an adaptable structure that can respond to changing water demands and dynamics, be based upon hydrological facts and usage rather than history or sovereignty, offer an equitable distribution of benefits and have well-defined and reasonable dispute-resolution mechanisms. It will not be an easy task, but it must be given a far higher level of priority than at present. Our international development policies would benefit from more debate on required outcomes rather than a constant discussion about the distribution of current spending.

The same improvements in our attention would be well directed towards increased access to clean water and proper sanitation for some of the world's poorest people. Not only would the reduction in the incidence of waterborne diseases bring enormous relief to millions of our fellow human beings, but it would also create a surge of economic potential that could enable some of the world's most impoverished to chart their own course out of poverty and disadvantage. If we want longer-term stability, then we must seek to eliminate the culture of dependence that exists in too many parts of the world. Providing access to clean water, decent sanitation and reduced levels of disease is key to this. It is surely an issue of global concern that, in a world where we can explore the limits of our solar system, there are still people, especially children, who die every day from lack of clean water. If there is one outstanding moral challenge to all of us in our current era, it is surely this. Where this pattern of expenditure occurs in countries which themselves have some of the worst water and sanitation issues, there are clearly issues surrounding their priorities to address too.

Finally, how do we deal with the whole issue of climate change and environmental degradation and assess its potential impact on all the issues we have considered. At the top end of the outcome spectrum, we might moderate our behaviour, find ways to satisfy consumer behaviour without polluting the natural world or degrading finite resources and limit or stop the global warming process. If we are to do this successfully, we should move away from lecturing our populations, particularly in the democratic world, and find solutions through innovation and technology that will enable us to achieve widespread co-operation without coercion. We cannot punish our populations into decarbonisation, damaging their quality of life or limiting their hope and expectation for the future. I believe this is well within our grasp if we give it the appropriate encouragement and support.

What does the worst-case outcome look like? We can go back to where we began this story, in a much earlier period of earth's history.

The following criteria were identified to sustain co-operative water agreements: (1.) an adaptable management structure that promotes flexibility in the face of changing water supplies, demands and values, (2.) clear and flexible criteria for water allocations based on water allocation schedules that adapt to extreme hydrological events, new information on river basins, hydrodynamics and changing social values, (3.) the equitable distribution of benefits that allows for positive-sum agreements and (4.) well-defined conflict resolution mechanisms.

How the water on planet earth has been distributed has varied over time and this dissemination has had profound effects on the pattern of life itself. It is worth noting that although the world has experienced periods of great heating and great cooling in our climate, mass extinctions have been more common during periods of rapid warming than of rapid cooling.

Around 250 million years ago there was a massive increase in the earth's volcanic activity which was centred on today's Siberia, with huge quantities of carbon dioxide and methane pouring into the atmosphere. Increased levels of acidity in the seas and oceans, caused by the absorption of carbon dioxide, are likely to have been toxic to many of the species which existed there. Worse, it is thought that changes to the ocean circulation itself also caused a reduction in the amount of oxygen in the water. This in turn led to the death of many marine species, including invertebrates such as coral, as well as many types of fish that were never to reappear. It is thought that the massive increase in greenhouse gases may have raised the average ocean temperature by as much as 8°C and it is believed that around 90 per cent of all species became extinct. Experts are divided over how long it took the world to recover from this catastrophe, with estimates ranging from 2 million years at the optimistic end of the scale to 10 million years at the more pessimistic. The survivors included the bivalves, the ancestors of today's mussels and scallops and the gastropods, relatives of today's snails and slugs. Eventually, these gave rise to the amphibians and reptiles who underwent their own period of global dominance and demise.

This is not to suggest that we face, as some of the climate pessimists maintain, a global cataclysm on this scale, but we must take care not to discard evidence we find inconvenient for whatever reason. We may be the most advanced species to ever dominate our planet, but we would do well to remember that our reign has been in the tens of thousands of years rather than the millions.

We have so much potential for good, improving the health and well-being of our fellow human beings and using our talent for creativity and innovation to protect an environmental balance that is more fragile than many care to acknowledge. We also, sadly, have a proven record of destruction and degradation.

Will the decisions we take lead us to the beneficial state of Jonas Salk, whose dreams ultimately conquered his nightmares? Or vice versa?

THE RIME OF THE ANCIENT MARINER

Samuel Taylor Coleridge spent his honeymoon in the beautiful seaside town of Clevedon, close to where I live in north Somerset. Clevedon has numerous associations with poets, including Alfred, Lord Tennyson who wrote *In Memoriam* there for his friend Arthur Hallam who lies buried in the grounds of St Andrew's Church. The church graveyard sits at the top of Poets' Walk, whose dramatic and picturesque path overlooks the Severn Estuary and commemorates both literary geniuses.

As Clevedon has no working harbour, it is thought that the inspiration for the port in the *Ancient Mariner* is Watchet in west Somerset, some 27 miles as the crow flies from Clevedon.

The poem is at once a story of the fickleness of the sea, a phantasmagorical narrative and a moral lesson. It is, in its tale of thoughtlessness, the abuse of nature and the recklessness of human ego, a tale for our times. On top of this we have echoes of Coleridge's own voyage into opium addiction and the graphic and towering images of the power of water in its various forms. It is easy to see why it is appreciated in such a myriad of ways. It is, along with Tennyson's *In Memoriam* itself, my own favourite poem.

The story is of a thin, grey-bearded and bedraggled sailor who uninvited, but compulsively, relates his own strange tale of a dark and mysterious voyage to an unknown guest outside a wedding. At first a reluctant and at times an alarmed audience, the guest becomes strangely captivated by the mariner's story.

Sometime after leaving port, the mariner's ship is driven south by a fierce storm and eventually reaches the icefields of the Antarctic. Here, an albatross appears, the presence of which is given a religious significance by the crew who feed and almost adopt the bird. For reasons that he does not, and probably cannot, explain, the mariner then shoots the albatross. The attitude of his fellow sailors changes variously from anger at the incident itself, through more benign feelings (though not forgiveness) as the weather improves and back to anger and blame when the ship is becalmed near the equator. As a sign of their anger and resentment, the albatross is hung around his neck, an act again imbued with religious symbolism.

> 'What evil looks
> Had I from old and young;
> Instead of the Cross the Albatross
> About my neck was hung.'

After a horrific period of dead calm and unforgiving thirst, a ghostly vessel approaches whose only occupants are a skeleton and a deathly pale woman, representing death and what the author describes as 'nightmare life in death'. The two play dice for the souls of the mariner and the crew, with 'life in death' winning the mariner and 'death' the crew. In the parched conditions, the crew die one by one, each fixing his last gaze on the mariner, emphasising his penance for his reckless act.

Eventually, the rain returns and miraculously, the crew come back to life and pilot the mariner back to his original port. Just before he is rescued by a small craft, the crew disappear into the hands of death once more and the ship sinks to the bottom of the sea in a whirlpool. The mariner is left to wander the land, telling his story, with all its implied lessons, to any strangers who will listen, including the wedding guest himself, who departs a sadder but wiser man.

I will leave to others the interpretations of the significance of events in Coleridge's own life and the wider moral and religious implications implicit in the text. In the context of this book, it is the dramatic images of the sea, in its various states, creating a breathtaking backdrop to the tale – sometimes threatening, sometimes optimistic – that I wish to highlight.

Coleridge's powerful images of water in its frozen and liquid forms permeate the narrative as he describes the voyage to the cold climes of the Antarctic with its ice shelves and icebergs.

> And now there came both Mist and Snow,
> And it grew wondrous cold;
> And Ice mast-high came floating by
> As green as Emerald.
> The Ice was here, the Ice was there,
> The Ice was all around:
> It crack'd and growl'd, and roar'd and howl'd—
> A wild and ceaseless sound.'

The fated albatross, before its sad demise, rescued them from their frozen environment.

> At length did cross an Albatross,
> Through the Fog it came;
> As if it had been a Christian Soul,

> We hail'd it in God's name.
> It ate the food it ne'er had eat,
> And round and round it flew.
> The ice did split with a thunder-fit;
> The helmsman steer'd us through!

When the ship is later becalmed, Coleridge gives a vivid description of both the conditions and the effects upon the sailors, particularly their crippling thirst.

> Day after day, day after day,
> We stuck, nor breath nor motion,
> As idle as a painted Ship
> Upon a painted Ocean.
> Water, water, every where
> And all the boards did shrink;
> Water, water, every where,
> Nor any drop to drink.

And again...

> And every tongue thro' utter drouth
> Was wither'd at the root;
> We could not speak no more than if
> We had been choked with soot.

> Ah wel-a-day! what evil looks
> Had I from old and young;
> Instead of the Cross the Albatross
> About my neck was hung.

As the ghost ship approaches, the full horror of their parching thirst is revealed:

> With throat unslack'd, with black lips bak'd
> We could nor laugh nor wail;
> Thro' utter drouth all dumb we stood
> Till I bit my arm and suck'd the blood,
> And cry'd, A sail! a sail!

After the dice game where the souls of the crew are divided, the long-awaited rains arrive.

> The silly buckets on the deck
> That had so long remain'd,
> I dreamt that they were fill'd with dew
> And when I awoke it rain'd.
> My lips were wet, my throat was cold,
> My garments all were dank;
> Sure I had drunken in my dreams
> And still my body drank.
> The thick black cloud was cleft, and still
> The Moon was at its side:
> Like waters shot from some high crag,
> The lightning fell with never a jag
> A river steep and wide.

The final sea imagery comes as the supernaturally crewed ship reaches the familiar port and sinks before the mariner is rescued.

> The Boat came closer to the Ship,

But I nor spake nor stirr'd!
The Boat came close beneath the Ship,
And strait a sound was heard!

Under the water it rumbled on,
Still louder and more dread:
It reach'd the Ship, it split the bay;
The Ship went down like lead.

Stunn'd by that loud and dreadful sound,
Which sky and ocean smote:
Like one that hath been seven days drown'd
My body lay afloat:
But, swift as dreams, myself I found
Within the Pilot's boat.

Upon the whirl, where sank the Ship,
The boat spun round and round:
And all was still, save that the hill
Was telling of the sound.

While it is the often quoted (and slightly misquoted) 'water, water every where, nor any drop (and not a drop) to drink' that is the most remembered line of the *Ancient Mariner* poem, the magnificent pictures generated by Coleridge's words are, to me, as stunning as any visual representation in art. Fog, snow, ice, thirst, rain, drowning. All the manifestations of water captured in perfect imagery.

ACKNOWLEDGEMENTS

Anyone who has ever written a book knows just what a team effort it really is.

Thanks to James, Lisa and all at Biteback Publishing for their help, support and faith in this project. My admiration for, and thanks to, all those whose academic, insightful and inspiring works provided the knowledge base from which I could draw to address the broad range of subjects herein. I hope I have reflected our mutual love of the subject matter. Thanks to all my wonderful staff: the legendary Ione, Ben, Joseph and Annabel for their steadfast support and willingness to listen to the same sections of the book being constantly rehearsed, to David for his help and loyalty over many years and to our wonderful Abraham Accords team, especially Freddie and Matthew.

To Sir John Major, whose idea it was to write this book in the first place, Professor Clive Finlayson for sharing the mysteries of the Gibraltar Neanderthals with us and to James Ashwood for his inspired cover artwork.

To all the friends who listened patiently while I shared and shaped my ideas: all our gang in Llafranc, James, Chris and Julie, Iain and Kenny, Mark and Ceri, Fran, Tom and Marie, to the wonderful 'Mayrhofen crew' Craig and Sue and Ellie for their continued encouragement and forbearance, and with love to Matthew and Alice for their future together.

To all my family, especially my brother Paul and my sisters Tricia and Louise, whose selfless love and devotion have kept so many different shows on the road and the young Fishers – Grace, Calum, Alexander and Anna – for caring about the health of the world around them. To Adam and Lucy for being the best friends in the world and for their limitless encouragement, loyalty and support and of course, my loving and ever-patient wife Jesme, who tolerates my obsessive projects with such elegant equanimity.

Apologies to all those too numerous to mention whose faith, friendship and support over the years have sustained me to this point, you have my permanent thanks.

Finally, to all those who work day and night to bring clean water and hope to all our fellow human beings – you have our perpetual gratitude and admiration.

INDEX